ADAPTABLE

ADAPTABLE

HOW YOUR UNIQUE BODY REALLY WORKS AND WHY OUR BIOLOGY UNITES US

HERMAN PONTZER, PhD

Avery
an imprint of Penguin Random House
New York

an imprint of Penguin Random House LLC
1745 Broadway, New York, NY 10019
penguinrandomhouse.com

Copyright © 2025 by Herman Pontzer
Penguin Random House values and supports copyright. Copyright fuels creativity, encourages diverse voices, promotes free speech, and creates a vibrant culture. Thank you for buying an authorized edition of this book and for complying with copyright laws by not reproducing, scanning, or distributing any part of it in any form without permission. You are supporting writers and allowing Penguin Random House to continue to publish books for every reader. Please note that no part of this book may be used or reproduced in any manner for the purpose of training artificial intelligence technologies or systems.

Most Avery books are available at special quantity discounts for bulk purchase for sales promotions, premiums, fund-raising, and educational needs. Special books or book excerpts also can be created to fit specific needs. For details, write SpecialMarkets@ penguinrandomhouse.com.

LIBRARY OF CONGRESS CATALOGING-IN-PUBLICATION DATA
has been applied for.

ISBN 9780593539309 (hardcover)
ISBN 9780593539316 (ebook)

Printed in the United States of America
1st Printing

Book design by Angie Boutin

The authorized representative in the EU for product safety and compliance is Penguin Random House Ireland, Morrison Chambers, 32 Nassau Street, Dublin D02 YH68, Ireland https://eu-contact.penguin.ie.

Neither the publisher nor the author is engaged in rendering professional advice or services to the individual reader. The ideas, procedures, and suggestions contained in this book are not intended as a substitute for consulting with your physician. All matters regarding your health require medical supervision. Neither the author nor the publisher shall be liable or responsible for any loss or damage allegedly arising from any information or suggestion in this book.

While the author has made every effort to provide accurate telephone numbers, Internet addresses, and other contact information at the time of publication, neither the publisher nor the author assumes any responsibility for errors, or for changes that occur after publication. Further, the publisher does not have any control over and does not assume any responsibility for author or third-party websites or their content.

For Janice, Clara, and Alex

CONTENTS

INTRODUCTION: DISASSEMBLY REQUIRED

"We are all ONE, say the gurus. Aye, I might agree—but one WHAT?"

—Edward Abbey, *A Voice Crying in the Wilderness*

In the vast Venn diagram of life experience, human dissection is that rare slice of real estate where the worlds of medical students and serial killers overlap. For one long New England autumn, my first semester of grad school, I deconstructed 160 pounds of person in my Harvard Medical School gross anatomy course. It was slippery and surreal work. Having grown up in a very "do it yourself" culture in rural northwest Pennsylvania, I was comfortable at an early age handling knives, hammers, and other hand tools, but usually on wood or metal. Never on *people*. Using a hacksaw to carefully bisect a human head, damp pale sawdust accumulating on the cheeks as the blade sunk down between the eyes, felt as though the Universe had made a category error, like living out some absurd dream where your aunt turns into a toothbrush or you find yourself driving a fish.

Like most non-psychopaths, my classmates and I had all started the course hesitant to cut into a human body. On day one, with all of us gathered around the lead instructor as he demonstrated the first, sweeping scalpel incision down the middle of a cadaver's chest, a student collapsed in a stress-induced seizure. I was an interloper, one of only a few in the class with no plans of becoming a doctor, and I expected some eager medical student (there is no other kind) to jump into action. Looks of alarm bounced around the room, but no one stepped forward. Seizure care isn't covered on the MCAT. I had spent the previous few years working on an ambulance crew and so I went to work, pulling her out from under the table, turning her on her side, keeping her head from banging on the floor. I didn't know her medical history, and was worried she might need serious medical attention. As she came to, trying to piece things together, I told her she'd had a seizure.

"Yeah . . . that happens," she said, a bit grumpy. Then she got up, dusted herself off, and asked what she missed. It turned out she was an interloper as well, an MIT biomedical engineering student taking the class to understand the inner world her inventions would someday inhabit. The seizures were a stress response she had learned to live with. We became dissecting partners.

People can adapt to just about anything. By the second week, the class settled into a rhythm, comfortable with the unglamorous work of disassembling humans. After a couple hours of lecture previewing the structures we'd be excavating, fifty first-year med students and a handful of biomedical engineers or biologists like myself would tramp upstairs to the dissecting theater, where a dozen bodies lay heavy and cold, zipped into thick black body bags on gleaming steel tables. Teams would assemble, donning white coats and latex gloves, four students per corpse. Over small talk, we'd unzip our cadaver's plastic cocoon and open our black pleather pouches of stainless steel dissecting tools, turn the greasy pages of *Grant's Dissector* to the day's menu, and dig in.

Our captain on this inner voyage was the unflappable Farish

Jenkins, PhD, a brilliant biologist and paleontologist famous for his work on ancient vertebrates' transition from sea to land and the evolution of mammals. Dressed impeccably in a starched shirt and tweed suit, his lectures had the urgency of a World War II newsreel. Farish would march us through every duct and vessel, the human body rendered in his impeccable hand through increasingly detailed multicolor chalkboard illustrations. He was as well-versed in the geography of the human body as he was gleefully eccentric. The lecture on gait, another one of his specialties, involved him clomping around the room with a wooden peg leg of his own design, reciting descriptions of Captain Ahab from *Moby-Dick*.

It was no accident that the lead instructor in Harvard's premier human anatomy course was an evolutionary biologist. An evolutionary perspective is essential for making sense of the human body. Farish's encyclopedic understanding of vertebrate evolution infused the course, illuminating the arcane and seemingly arbitrary details of our anatomy. (It's nonsensical that lungs and stomachs should develop from the same group of cells in an embryo, but obvious once you know how lungs evolved from swim bladders, outgrowths of the fish digestive system.) The other instructors were an all-star cast of doctors, biomedical engineers, and experts in the art of dissecting humans.

Even with all this powerhouse instruction, most weeks (if not most days) found my fellow students and I adrift, poking about our cadavers in a fruitless search for the structures we were meant to find. The web of nerves spraying out from the base of the neck to control the arm and fingers, the delicate weave of muscle and connective tissue in the pelvic floor . . . Sometimes things were precisely as depicted in Farish's drawings and our textbooks, but often they were not. These weren't minor anatomical details that could safely be ignored. These were critical structures, like the branching vessels that leave the descending aorta to feed the abdominal organs, or the vagus nerve that controls heart rate (it was stress-induced stimulation of the vagus nerve that occasionally sent my dissecting partner into a seizure). Roving instructors were kept busy each afternoon, moving from table to table, hunting down lost structures for flummoxed students.

All of this perplexing variation was the point. People who donate their bodies to science skew White, wealthy, and old. And yet, hidden within the seemingly homogenous slice of humanity laid out on our dissection tables was a surprising amount of biological diversity—individual anatomies we'd never find in our textbooks. And this was just the variation we could see with our naked eyes and amateur dissecting skills. More diversity lay unseen beneath the surface, within the tissues, deep inside the cells themselves, down to their DNA. All the illustrations and *Grant's Dissector* diagrams were a convenient fiction. There is no textbook human.

When these people were alive, their bodies in motion, these differences multiplied into the complex physiologies that made each person unique. High blood pressure, a fondness for ice cream, deeply felt introversion, a lifelong struggle with the bathroom scale, a knack for distance running. Fragments of these stories could be pieced together from telltale clues: an implanted pacemaker discovered next to the heart; an artificial hip; limbs lean and muscular or sagging with fat. I suppose it shouldn't have been a surprise. Our differences are obvious, even on the surface. Why should our insides be any less diverse?

Any course attempting to navigate the entirety of the human body in one semester is forced to balance all this variation against the seemingly contradictory reality that we're also all alike. We've each got a brain in our head, a heart in our chest, and a digestive tract running end to end. The textbook illustrators play the odds.

For any particular piece of the body, most people look pretty similar. Chances are good that your vagus nerve travels the textbook route, even if the probability that *every* part of you matches the diagram is essentially zero. And much of the variation we experience in our daily lives, including the health issues the medical students would encounter as doctors, comes from subtle differences in how our parts behave or from our environments, not obvious differences in anatomy. The effectiveness of a liver or heart isn't usually apparent from its outward appearance. My dissecting partner's seizures couldn't be discerned from the path of her vagus nerve. In a class focused on the quiet anatomy of still bodies, the foundational course for aspiring doctors, we acknowledged the diversity but studied the textbook.

There's another reason that courses on the human body tend to sidestep the question of diversity. Differences are dangerous. Focusing on the ways that people differ from one another has an ugly, racist history in biology, one that doctors and researchers are eager to move past. Even today, some of the most cutting-edge research exploring the genetics and physiology that makes each of us unique also makes many people uncomfortable. The old "nature versus nurture" debates emerge. There is understandable concern about giving fresh fodder to old, discredited racist, sexist, and eugenic ideas. Discussing differences pulls people apart.

But ignoring our biological diversity only makes it more challenging to grapple with. Thorny concepts like race and ancestry, or gender and sex, begin to feel impenetrable, even for doctors—the people trained and entrusted to understand us completely. Medical professionals, who presumably should know better, fall back on the same clumsy categories we all use, and share many of the widespread and unfounded biases about how people and their bodies differ, with real consequences. In the U.S., your doctor's perception of your race can affect everything from the amount of medication they prescribe for pain relief to their assessment of whether you are overweight. Many of these biases are baked into algorithms that doctors rely on every day—seemingly objective equations that calculate a patient's health status and guide medical decisions. The exact same lab results from a urine test, churned through the standard equations for kidney function, will flag a White patient for treatment, but indicate a Black patient is fine. Other race-based calculations affect medical care for heart disease, cancer, hip fractures, and even how a mother delivers her baby. The evidence behind these race "adjustments" is typically slim or nonexistent, and inevitably conflates the effects of environment and socioeconomics with underlying biological variation. Assumptions based on sex, gender, and age are often equally tenuous.

The rest of us, without any formal education in our biology, are left even deeper in the dark, vulnerable to social media grifters and self-help charlatans. If you don't know where your kidneys are or what they do, it's easy to be sold on the latest detox product, or believe that your skin color affects the way your kidneys work. If you don't know how your immune system functions, you can be duped by the latest anti-vax propaganda. If you don't know how blood types work, you can be conned into thinking they determine the diet that suits you best.

There are societal reverberations as well. The way we understand ourselves determines the way we see others. If you don't know why some people's skin is darker, it might seem reasonable to think that race categories represent deep, biological, and inescapable differences between groups (they don't). And there's a personal cost to our anatomical ignorance. It robs us of a deep appreciation of the wondrous vessel we inhabit during our time here on earth.

This book presents a different way to think about the human

body. Like my gross anatomy course, we'll take a tour of the miraculous protein robot that is your physical self. And, like Farish, we'll make sense of the human body using an evolutionary lens. But crucially, when it comes to understanding our diversity, we will make the other choice, moving beyond the textbook average human and focusing on how and why we differ, and why you are unique. Drawing on advances in genetics and insights from populations around the globe, we'll develop a new, more complete picture of ourselves and our diversity. As we'll see, far from pulling us apart, our diversity unites us.

WHY OUR BIOLOGY MATTERS

Human diversity is endlessly fascinating, especially for biologists like myself who study how our evolutionary history shapes our anatomy and physiology. I've had the good fortune to spend my career investigating our evolutionary past and our diversity today, excavating fossils and stone tools in the Republic of Georgia, living with traditional hunter-gatherer and pastoralist communities in Tanzania and Kenya, and investigating health and metabolism in my lab at Duke University here in the U.S. Colleagues and collaborators have pushed even further afield, working with farmers, foragers, and fishermen around the globe to understand the connections between lifestyle and health. A revolution in genetics and portable, field-ready technologies has thrown the doors of discovery wide open, bringing surprising new insights and challenging old ideas about how our bodies work. The science of the human body has never been more exciting.

It's a wonder, then, how the public discussion of these life-changing advances has been so limited. The little discussion we've had has been stuck in old tropes about "nature versus nurture," or built on outdated ideas. Many, including leaders in public policy and academia, have been slow to embrace and engage with the burgeoning science of human diversity.

Their hesitation is understandable. There is a shameful history of biologists promoting racist, sexist, and other exclusionary and oppressive ideas, including slavery. Work by prominent biologists in the 1800s and early 1900s contributed directly to eugenic policies and laws that stripped people of their rights and personhood, culminating in the horrors of the Holocaust in World War II. The collective revulsion to those atrocities helped evict so-called race science from mainstream biology, but one doesn't need to look very hard to find racist arguments alive and well in the public discourse, often with a pseudoscientific veneer.

The reformers who pushed back against the racists throughout the 1900s did heroic work, fighting against the scientific consensus of the time to establish that race was a cultural convention, not a biological reality. Modern genetics research, including the analysis of DNA recovered from skeletons at archaeological sites, has only strengthened this perspective. Humans are a relatively young species with our origins in Africa, and we've been migrating and intermarrying from the beginning. Our gene pool is so shallow and so thoroughly mixed that every one of us all over the world is more than 99.9 percent similar in our DNA. When we look at particular genetic variants (called alleles), like the kind that produce type A versus type B blood, we find that any two populations from anywhere on the planet share over 90 percent of their variants. The genetic differences that set us apart are the chop and foam atop an ocean of similarity, and they don't cleanly separate us by race.

By the middle of the twentieth century, the reform movement had grown beyond concerns about race to include sex, gender, sexual orientation, and other modes of diversity. Many rejected any suggestion that inborn biological differences had a meaningful influence on who we are or how we differ, a view that became canon across much of the social sciences. Author Kurt Vonnegut summed up the prevailing mood nicely in his 1969 surrealist classic, *Slaughterhouse-Five*: "I went to the University of Chicago for a while after the Second World War. I was a student in the Department of Anthropology. At that time, they were teaching that there was absolutely no difference between anybody. They may be teaching that still."

Indeed, they are. Today, research into biological variation among people or populations is often met with suspicion, perhaps even whispers of racism, sexism, or other nefarious intent. In the past couple of decades, we've seen anthropology departments at Harvard University and elsewhere torn apart as people who see humans as purely cultural beings find little common ground with those like me who are also interested in our biology.

Again, it's an understandable perspective given the history of bigotry and oppression in biological science. But an absolutist position, that environment and culture are the *only* meaningful factors shaping our bodies and minds, seems increasingly out of step with the growing evidence that our biology, including our genes, matters, too. DNA differences don't cleave along race lines, but that doesn't mean they aren't important for particular populations or meaningful for individuals. And I would argue that ignoring genetic and other biological differences cedes too much of the public discourse to the racists, sexists, and other jerks. It's natural for people to notice differences among individuals and groups and wonder where those differences come from. Humans are inherently curious. We want answers and deserve the best science.

We also need to acknowledge and understand our biology in order to be informed citizens. From the COVID pandemic and discussions about vaccines and masks, to the endless debates about abortion and the beginning of life, to age-old questions about racial differences, to the more recent debates surrounding sexual orientation and gender, the ideas and disputes that dominate the public discourse today all have roots in our biology. We need to understand how bodies work and why they differ if we hope to find common ground, make evidence-based decisions, or even to fully articulate our own perspectives.

Finally, ignoring our biology robs us of a full understanding of ourselves. Two decades of teaching college physiology has left me convinced that most adults know more about the social lives, sexual proclivities, and health concerns of celebrities than they do about their own bodies. There's a curious double standard about what a well-rounded adult should know. We roll our eyes at people who can't operate their smartphones or don't get a Shakespeare reference, but shrug at our collective biological ignorance. You will spend your entire life inside your body, and a working knowledge of its parts and their functions will serve you well. It can help to keep you healthy, for a start. And it can help make sense of the big questions. *Why is my body like this? How does life start, and why do we die? Why are we so alike, yet so different? What makes me, ME?*

A book about our bodies is inherently more than just a textbook. It's about life, identity, health, our relationship with others, and what it means to be human. I wrote this book with two audiences in mind, and you are both of them: the person interested in how your body works and why you're unique, and the person embedded in their community, making decisions about how we treat one another. I hope you'll use this book to engage with your self and your fellow humans, shout down racists, make sense of the latest headline in science and medicine, consider the ancestry you inherit and the environment you inhabit, take part in public debate, talk to your doctor, and ponder more deeply this human life and your strange, wonderful, self-made human body.

ADAPTABLE

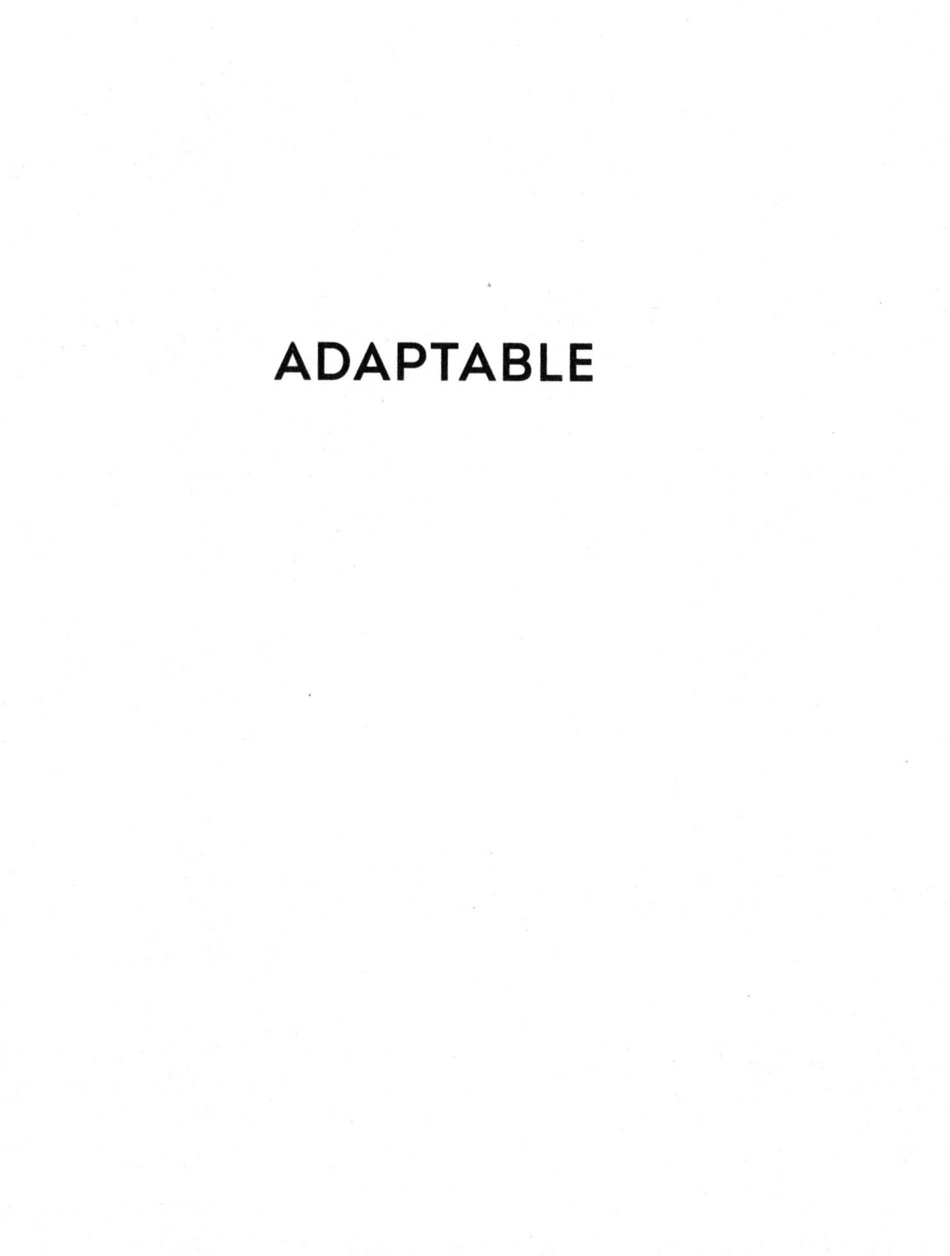

1

SO SIMPLE A BEGINNING

How old are you?

If you're reading this book the answer probably comes easily. Your birthday was one of the first dates you ever committed to memory, the promise of presents and cake cementing the coordinates into your internal calendar. As an adult that day might be less meaningful, or even cause for despair. But you know it. You know *exactly* how old you are.

It might surprise you, then, that many people around the world today don't have a precise idea of their age. Much of my research takes me to communities like these, that don't keep calendars or track time the way we do in the industrialized world. It was a bit of a shock the first time I encountered it.

I was living and working with the Hadza community in the dry, wide-open savanna of northern Tanzania. It's a small and tight-knit population. Only about 1,200 people consider themselves Hadza and speak the language. The Hadza are hunter-gatherers, living in small grass houses clustered in camps among the acacia and baobab trees, making their living off of the wild landscape around them. Every morning, women leave camp in groups, on foot, to forage for berries or wild tubers. Men head out, usually alone, to hunt for antelope, giraffe, and other wild game using bows and arrows they make themselves. In a world that has rapidly moved to farms, then towns, and then to industrialized cities, the Hadza have maintained their traditional way of life for generations, perhaps millennia. There are no domesticated crops or animals in a Hadza camp, no electricity or plumbing, no guns or cars. And no calendars.

Settling into camp life on my first trip there, I would see kids running around, hanging out with mom and dad, sharing a meal around the fire. Trying to make a bit of friendly conversation, with the best Swahili (our common language) that I could manage, I'd ask a mom or dad how old their son or daughter was. The usual answer was a shrug.

Who doesn't know how old their kids are? I would wonder. Then I encountered it again when we interviewed the adults and asked their age. *No one knew.* Our research team spent days each season working to figure out everyone's age.

It's a common theme in fieldwork with nonindustrial communities. I've encountered the same issue working with traditional pastoralists in northern Kenya, and my collaborators have had similar experiences all over the world. Accurate calendars that track our planet's orbit around the sun are a relatively recent invention of farming cultures that needed to know precisely when to plant their crops and when to harvest and prepare for winter. Lunar calendars are more common, and potentially more meaningful. Their cycle seems more in tune to the rhythms of daily life and our attention spans, and they determine how light the nights are in cultures without electricity. We've even kept them embedded in our Western calendars, though we've stretched them into months.

It's easy to be a bit smug about the tools we have to track time, as though by measuring it accurately we might exert some control. But do you *really* know how old you are, even with your calendar? You know when you were born, of course, but what about all the stuff that happened before that? The day you were born, your body had already been under construction for months. In some ways, it's been under construction for more than a billion years.

ENDLESS FORMS MOST BEAUTIFUL

Watching the body develop is a bit like taking a walking tour of the history of life. You watch a single cell become a multicellular cluster, then fold itself into a simple tube with a gut and basic nervous system, grow a head with eyes and a mouth, and sprout limbs. We see vestiges of ancient structures as the embryo develops, hints of old designs that were once the norm but have since been abandoned or changed beyond recognition. A yolk sac forms in the first few weeks only to dissolve away. The jaw, ears, and throat grow from fleshy arches that would have become the gills of our fishy ancestors.

There's an old-fashioned idea that we pass through each stage of evolution as we develop. "Ontogeny recapitulates phylogeny" is how I learned it. That's not quite right—you don't pass through a fish stage, or a reptile stage. But the layered nature of evolution, each generation tweaking and adding to what came before, is obvious from our embryos.

About nine months before birth, somewhere near the outer reaches of one of mom's fallopian tubes, a sperm breaks through the wall of a recently ovulated egg, fertilizing it. (If you're not clear on how we get to this step, you might check out chapter 7 or schedule an awkward talk with your parents.) DNA carried by the sperm and egg, packaged in tightly wrapped chromosomes like skeins of yarn, will combine to become the embryo's genome, unique among all life, with the recipe for a human written in a four-letter code some 3 billion units long. After some gymnastics to package the twenty-three chromosomes from the sperm and the twenty-three from the egg into a single forty-six chromosome nucleus, it's time to grow.

That one cell divides into two, which each divide to make four, then eight, sixteen, thirty-two . . . all the while floating lazily along the fallopian tube, toward the uterus. At around sixty cells big, the cluster of cells fills with fluid, like a new soccer ball that somehow learned to inflate itself. Around five days after fertilization this fluid-filled ball of cells lands upon the uterus surface and burrows in, implanting. The cells embedded in uterus will grow into the placenta, nestling up against the blood-rich tissue of the uterus to pull in oxygen and nutrients. Only a tiny clump of cells on the inner wall will grow into the body.

At about three weeks after fertilization, that clump of cells grows and organizes itself into three distinct layers (picture a stack of three slices of bread). This three-layer sandwich curls to form a roll, and the layers form concentric tubes. The bottom layer becomes the innermost tube, the top layer becomes the outer tube, and the middle layer rests between them. That inner tube will become your digestive tract, from mouth to butt. And that's the basic body plan for fish, birds, reptiles, and mammals like us. We are all just elaborate tubes.

At the same time your main body tube is forming, a second tube forms down the middle of the outer layer. A deep groove develops along the surface, marking the track that will become your spine. The sides of the groove rise and crash into one another like cresting waves. They fuse together where they meet, converting the open gully into a closed "neural tube." As the name suggests, the brain and spinal cord will fill the neural tube eventually. Production of neurons, the cells that store and transmit information, begins in the sixth week after fertilization and continues throughout gestation.

The rest of the outer layer will develop into skin and hair as well as the skull. The middle layer will develop into the muscles, heart, and blood vessels, most of the skeleton, kidneys, and reproductive organs. The stomach, intestines, liver, and bladder all form from the innermost layer, along with the lungs. Timelines for organ development vary. Some get started in the fourth week after fertilization, and many continue to develop through gestation and into childhood. By ten weeks of development, things are far enough along that we switch from calling the growing organism an "embryo" to calling it a "fetus."

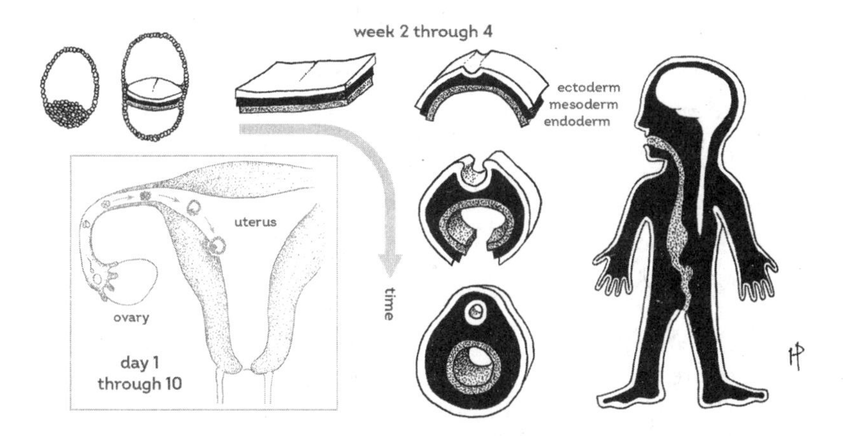

Figure 1.1 Early embryo development. Fertilization occurs in the fallopian tube, near the ovary. The embryo then divides repeatedly as it drifts toward the uterus, eventually implanting in the uterine wall around day nine. The clump of cells within the early embryo organizes itself into three distinct layers, the ectoderm (white), mesoderm (black), and endoderm (gray). This three-layer sandwich folds to form the gut tube, which will develop into the digestive tract and lungs. The ectoderm forms the neural tube, which will develop into the brain and spinal cord. As a result, your skin, brain, and spinal cord develop from the ectoderm layer, your digestive tract and lungs from the endoderm, and your muscles, bones, and most other tissues from the mesoderm.

The origami of early development is tricky, and there is little tolerance for error—so little, in fact, that all animals use a nearly identical set of patterning genes to get started. That's why an early chicken embryo is hard to distinguish from that of a human. It took untold eons for these genes to evolve, but once in place they became so fundamental to development that they've barely changed for half a billion years. Every once in a while, a mutation in one of these early patterning genes turns out to be beneficial, leading to the evolution of new forms. But much, much more often, tweaking these early patterning genes ends in disaster for the developing embryo.

Other dangers lurk as well. An organ or a limb might not form properly. Hormonal signaling with the mother might break down, causing the embryo to be spontaneously aborted. If the neural tube doesn't close all the way, the baby will be born with a hole through its back or neck that, depending on the location, exposes the spinal cord or brain. Modern medicine has made important strides in reducing the risks of some of these problems. For example, we know now that taking folic acid (folate) supplements before and during pregnancy reduces the risk of neural tube defects like spina bifida. But even for healthy mothers in countries with top-notch health care, the rate of loss is remarkably high. Roughly 30 percent of embryos are spontaneously aborted, and about half of those are lost very early, before a woman is aware she's pregnant. You're lucky to be alive.

WHEN DO WE START?

So when, exactly, does a life begin in the midst of all this folding and dividing? As we'll see throughout this book, the way we understand our bodies informs the way we understand our world, and there may be no greater concern than life's origins. Even if we want to remove ourselves from the heated, personal arguments surrounding abortion, it seems we ought to have some clear idea of when life starts. It's the ultimate existential question. The answer is not at all obvious.

From a strictly biological perspective all cells are equally alive (unless they're dead), chugging along, carrying out their daily metabolic processes. One of those processes is cell division, one cell splitting into two, the fundamental process of reproduction. Your life is the continuation of an unbroken chain of living and dividing cells that began billions of years ago. You could claim any point in the history of life as your personal origin. From that perspective, there's no clear point where *you* begin.

We can say life starts at fertilization, but that's still arbitrary. Eggs are roughly twenty thousand times larger by volume than sperm cells, which means we're more than 99.99 percent egg at the moment of conception. Adding a sperm to an egg doesn't reset its age any more than hanging a new air freshener from the rear-view mirror of my 2004 Honda Civic makes it a new car. The egg from which you developed was produced while your mother was still an embryo, just nine weeks into development. It was already very much alive and probably decades old by the time it was fertilized. Your single-celled embryo was equally old, at least all but 0.005 percent of it.

If we want to know when someone's life begins, a better question might be *When does a developing embryo or fetus become a person?* Biology is less equipped to answer that one. There's no biological definition of personhood—we've crossed into the realm of law, philosophy, or religion. Still, our legalistic or philosophical definitions would presumably be based on some biological criteria. Is there a clear point at which we can draw a line, one that doesn't have fuzzy edges and strange implications? As a biologist, I'm not sure there is.

There's the "life begins at conception" view, that fertilized eggs and very early-stage embryos should be treated as people because, unlike sperm and unfertilized eggs, they have the ability to become a unique individual. This perspective solves the issue of old eggs, but it raises other questions. First, how do we classify "monozygotic," or identical, twins, in which two individuals develop from the same fertilized egg? Surely a single fertilized egg can't be *two* people, but then one of the twin's lives (or both of them, depending on your perspective) must have started when the early embryo divided, not at conception. Perhaps we could say that *any* early embryonic cell or cell cluster with the potential to develop into an individual is a person. That would solve one twin problem but raise another. How would we classify conjoined twins, in which a single embryo only partially divides—one person or two? And how would we classify chimeras, in which two separate early embryos fuse together, resulting in one human?

Considering an embryo a person also raises philosophical questions about the hundreds of millions of spontaneously aborted early embryos, and more pragmatic issues for fertility clinics that create and store embryos for in vitro fertilization (IVF). IVF creates several embryos in test tubes, only some of which are implanted to start a pregnancy. The others are frozen or discarded. Should that practice be outlawed? That question holds stark implications for people relying on IVF to start a family. A 2024 Alabama Supreme Court decision that argued a fertilized egg is a person shut down fertility clinics across the state.

The view that personhood starts with fertilization is also complicated by modern technological advances that have removed the need for fertilization. We have the technology now to convert nearly *any* cell in our body into an embryonic cell with the capability to become a unique individual. It's not legal to do this with human cells, but that doesn't mean it won't be done. Would that embryonic cell qualify as a person, even though it's not the product of fertilization? What about the adult cell it came from? It's true that turning an adult cell into a developing embryo requires a lot of outside intervention, but the same is true for a fertilized cell in the womb. No embryo can develop entirely on its own.

Another common argument against the "life begins at conception" perspective is that a cell doesn't have a brain and nervous system. If we reserve the definition of a person to an entity that is able to perceive the world and have at least some limited sense of self, then a fertilized egg or early embryo is no more a person than a clump of liver cells would be. But brain-based definitions of personhood make people queasy, for understandable reasons. Newborn infants have limited awareness compared to adults. People with brain injuries, developmental problems, or dementia might have severely compromised cognition. If you're in a coma or under anesthesia, your conscious brain activity might effectively be zero. Do we limit or deny personhood in these cases? In some ways we do: parents exert a lot of control over their children's lives, and adults with severe cognitive disabilities often have limited autonomy. But we still recognize all of these individuals as people, deserving of dignity and essential human rights.

Those who believe that life begins at conception argue that a brain-based definition is a slippery slope that, taken to its logical conclusion, would deny the personhood of anyone with mental limitations, including infants. And if we deny those individuals personhood, the argument goes, then killing them would be morally equivalent to killing a farm animal or a fly. This isn't just fearmongering from religious conservatives or antiabortion activists. Noted philosopher and atheist Peter Singer has famously argued that killing newborns isn't equivalent to murdering adults because infants lack a sense of the future and therefore aren't people. That's a bridge too far for many (certainly it is for me).

I don't believe we need to abandon brain-based definitions altogether, however. And I don't think most "life begins at conception" folks really would, either. Consider two cases in which an embryo partially divides early in development, while it's still a clump of undifferentiated cells. The two cases are equivalent in every aspect except that in one the split occurs at the head end, while in the other it occurs at the tail end. In the embryo where the division occurs at the head end, two heads develop that share one torso and set of limbs. For the other, two abdomens, pelvises, and sets of legs develop, but just one head. These situations are rare, but they both occur in humans, and we treat them very differently. The first, with two heads, is considered to be two people who happen to be conjoined twins. The second is considered to be a single person with four legs. The distinction, in terms of the number of people represented, comes down to the number of brains. Ignoring brains as a criterion for personhood would require us to treat both cases as either one person or two.

There's an even rarer condition that helps to clarify the importance of the brain for personhood. In craniopagus parasiticus, which occurs less than once per million births, a baby is born with a second head attached atop its own. The second head has a brain, but with minimal or no function. Is the second head a person? In the dozen or so cases ever documented, there's at least one in which the second head was given a name. That case is notable because the second head had some minimal signs of autonomy; it could smile and blink. Usually, these babies are considered to be one person with a severe developmental condition rather than two people, consistent with the lack of apparent consciousness in the second head. Not only do you need a brain to be considered an individual person, the brain needs to meet some basic, functional threshold.

But defining personhood based on the development and function of a brain or any other structure is still fraught. Organs develop bit by bit, cell by cell, over days, weeks, and months. The heart develops earliest, and the pulsing of its early, rudimentary chambers can be detected as early as six weeks after fertilization. At that point the long development of the brain is just beginning, and unlike the heart there's no definitive, telltale signal that it's online and active. Electrical activity in early neurons can often be detected by the seventh week, while the connections between neurons needed for normal brain function don't develop much before twenty-two weeks. The fuzzy point at which brain function crosses over into something we'd recognize as human is currently anyone's guess, and depends on the threshold we set for being human.

What if we reserve personhood for a fetus able to survive on its own, outside the womb? The vital organs aren't developed enough for that until about twenty-four weeks, and only then with a lot of modern medical care. But that definition of personhood is dependent on the available medical technology, and that can differ widely across time and location. Why should a twenty-four-week fetus with access to modern medicine have a different personhood status than one without it?

Our definition of a person isn't the only issue at play in the abortion debate. Some argue it's not even the central question, that instead the critical issue is whether one person (the mother) should be required to use their body as a life-support system for another (the embryo or fetus). They would note that we don't generally make people use their bodies to save or support the life of another. We don't require people to donate blood, even though donated blood saves lives. Nor do we mandate that everyone be a living kidney donor (you only need one) or liver donor (they only take a portion, and it grows back), even though thousands die in the U.S. each year waiting for transplants.

One could argue that pregnancy is different, assuming the mother had consensual sex and knew pregnancy could result. I'm not sure that's fair in the case of failed contraception that's usually highly effective; both the man and woman had a reasonable expectation that a pregnancy wouldn't occur. (This also raises the question of our expectations and requirements of fathers.) And it wouldn't apply in cases of rape, including situations where the woman was too young, intoxicated, or otherwise unable to give consent.

It also doesn't account for people changing their minds. If you agree, in a moment of intense emotion, to donate a kidney to someone dying of renal failure, you're allowed to change your mind in the U.S., even if it means the person will die. Perhaps you're diagnosed with cancer and need to start treatment right away, or you realize the donation would put you at heightened risk because of your age. Maybe you receive devastating news that the recipient has already died, or was so ill that the transplant would only guarantee a few painful months of existence. Or you simply change your mind. You're allowed, because it's your body and no one can tell you what to do with it.

THE OUTSIDE WORLD SEEPS IN

Abortion may be the starkest example of outside interference in development, but it's far from the only external factor that affects a pregnancy. When we discuss embryological development, it can seem as though the entire process runs on autopilot, like some NASA satellite preprogrammed to unfold its delicate instruments alone in the dark, with only its hardwired computer code to guide it. While the guidance of our genetic instructions is clear, signals from the environment help shift and shape development in the womb as well.

The idea that a mother's behaviors and experiences affect the developing baby is old and universal. Every culture has its own list of dos and don'ts during pregnancy. Traditional Chinese taboos during pregnancy include prohibitions against mutton (the baby might develop epilepsy), watermelon and mung beans (these are "cold" foods believed to slow the circulation), and using scissors while on a bed (could cause cleft palate). Zulu women are advised to avoid eating eggs or liver (the baby will be born without hair). Women in Ghana are advised to avoid eggs as well, lest the child develop into a thief. A persistent myth among U.S. mothers warns that raising your hands above your head can choke the fetus with the umbilical cord.

Many of these traditions are harmless, the product of generations of women doing their best to understand what helps or hurts, seeing patterns when nothing is really there. Some, like the taboo against lifting your arms, seem to be the product of an overactive imagination. Others could do more harm than good, taking nutritious foods off the menu or restricting other healthy behaviors. And some, unlike those listed above, are undoubtedly helpful. Avoiding alcohol and extremely stressful situations, common themes of traditional pregnancy prohibitions, is probably a good idea for everyone, pregnant or not.

Modern science has worked to move beyond folk wisdom to identify environmental exposures that can affect fetal development. The big findings are well known and probably not terribly surprising. Alcohol, nicotine, opioids, and other drugs can have devastating effects on the developing baby. These molecules cross from the mother's bloodstream through the placenta and into the fetal bloodstream. There, they can cause all sorts of damage to the developing structures, affecting everything from fetal growth to brain development. (You can still drink coffee and tea, by the way—there's no evidence for any harm from a cup or so per day.)

It seems obvious that poisons and pollutants can harm a pregnancy. More surprising has been the evidence linking maternal nutrition and stress to fetal development. These effects can stay with the fetus through birth and infancy and into adulthood. In fact, emerging evidence suggests they can persist into future generations.

The Dutch Famine of 1944–45, also known as the Hunger Winter, was one of the many manufactured tragedies of World War II. In the early autumn of 1944, as revenge for the Dutch resistance conspiring with the Allied invasion of Europe, the Nazis cut off nearly all of the Netherlands from shipments of food. By October, food shortages were common throughout the country, particularly in the northern cities. By January, the situation was truly desperate. In all, about twenty thousand people died from starvation that winter and spring, before food supplies could be reestablished.

In the aftermath of the war, as people everywhere came to grasp the scale of its many horrors, researchers in public health sought a better understanding of starvation and its effects. Some turned to the Dutch Famine as a kind of natural experiment. They were particularly interested in the effects of famine on pregnancy, since the energy demands of fetal development are so high. A typical pregnancy for a healthy woman in a well-nourished population requires over seventy-five thousand kilocalories. How was development affected when those calories simply weren't available?

Medical records of mothers who were pregnant during the famine revealed that they gained less weight, just as we might expect. The effect was particularly strong among mothers who experienced the food shortage late in pregnancy, when weight gain is typically the fastest. Normally, mothers can expect to gain four to five kilograms (about ten pounds) during the third trimester. Women in their third trimester during the Hunger Winter gained *nothing*. And their babies were smaller, too, weighing around 5 to 10 percent less at birth than babies born to Dutch mothers before the famine.

Babies born in the Dutch Famine didn't stay small. Infants and children have a remarkable capacity to make up for lost ground, gaining weight rapidly when food is finally available, to catch up to their well-nourished peers. But as we've come to learn, that doesn't mean those early deficits are forgotten. In the 1980s, the epidemiologist David Barker published landmark studies tracking birth and death records in Wales and England. He and his colleagues showed that people born with low birthweight in the early 1900s had a much greater risk of heart disease and dying from a heart attack as adults in the 1970s. He proposed a radical idea that poor nutrition in the womb leads to changes in development that affect health and mortality into adulthood. The Barker hypothesis, as it became known, has been supported again and again in subsequent studies. Malnourished newborns in populations around the globe are more likely to develop heart disease, diabetes, and other problems as adults, even when we control for other factors like income and age.

We see this effect with Dutch Famine babies as well. Decades after the war, when they reached their fifties, adults who had gestated during the Hunger Winter were more likely to have coronary heart disease and elevated blood sugar (an early sign of diabetes) than those born just before or conceived just after the famine. They also tended to weigh more and report more anxiety and depression. By the time they reached their sixties, more Dutch Famine babies were dying of heart disease and cancer, and those still living showed more signs of cognitive and physical decline. The implications for all of us are clear: We never truly forget our past. Early life experiences stay embedded in our bodies for a lifetime.

Researchers are still trying to pin down precisely how early exposures, even in the womb, persist into adulthood. Current thinking focuses on two effects. First, without sufficient calories to support growth in utero or early childhood, the organs don't grow as large or develop as well as they would with better nutrition. These deficits last a lifetime and could cause problems as the body begins the usual age-related decline later in life. Second, poor nutrition in early life seems to change the way the body handles calories when food becomes available. Low birthweight babies tend to put on weight rapidly, particularly body fat. It's as though the body learns to expect that food will be hard to come by, and reacts by storing as many extra calories as possible in case starvation comes again. As a result, paradoxically, being born *underweight* is a risk factor for being *overweight* in late childhood and adulthood.

New, exciting research is beginning to show just how these changes occur, uncovering genetic modifications in response to poor nutrition. The growing fetus can't change its DNA, but there are ways to turn genes off or make others more active. These "epigenetic" changes to our DNA have major effects on how the body functions. When they occur during critical developmental periods, such as gestation and early childhood, they seem to persist into later life, affecting our risk of disease as adults. In the Dutch Famine babies, epigenetic changes to their DNA, in genes that affect weight and inflammation, were still evident in their sixties.

Even more remarkably, these effects seem to echo into future generations. Children of the Dutch Famine babies (the grandchildren of the women who were pregnant during the Hunger Winter), now in their thirties, weigh more and have a higher body mass index (BMI) than their peers. And it doesn't matter if it's the father or the mother who was exposed to the Dutch Famine prenatally; these transgenerational effects are passed down through both. Since fathers provide little more than chromosomes to the developing embryo, the clear implication is that the effect is passed down through epigenetic changes in the DNA.

Other multigenerational studies have pushed these results even further, indicating that both men and women exposed to different nutritional regimes as babies can pass those effects to their children and *grandchildren*. Studies like these are hard to conduct in humans because they rely on data collected across generations, and old records are often incomplete. Lab research on transgenerational effects with mice and other fast-reproducing species clearly shows these epigenetic changes persist across generations. While the verdict is still out on how long, exactly, these effects can last in humans, there's growing evidence that our DNA holds the experiences of our ancestors. The environments that your parents (and perhaps even your grandparents) experienced are imprinted in your DNA.

Evidence for the lifelong effects of early environments should give us pause when we assess the landscape of human diversity. Health disparities among adults today are not only influenced by their genes and lifestyles but by the environments they experienced decades ago in the womb and as young children. For older Black Americans and other minority groups, those early environments predate the civil rights era. Even today, nutritious food and health care remain out of reach for too many of us, and the burden falls heaviest on minorities and the poor. We would like to think that fixing these problems wipes the slate clean, but it's not that simple. Babies born today might still carry the epigenetic insults of their grandparents. The body remembers.

SO HOW OLD ARE YOU?

Your body is a work in progress, a mosaic of old and new. The genes that shaped you as an early embryo are hundreds of millions of years old, the DNA sequences handed down faithfully over eons like sacred texts. The genes that come on later, to build you into something human-shaped instead of fish- or ape-shaped, are much younger, but still millions of years old. Your unique genome, the DNA sequence that defines you, is cobbled together from bits of chromosomes that we could trace back through the generations of your family tree. You carry mutations in your genome, new sequences unique to you that neither of your biological parents carry in their DNA. There are modifications in your DNA, little epigenetic notes, marked by the environments your parents and grandparents experienced, like fingerprints pushed into wet clay.

Your own embryological origami began nine months before your birthday, from an egg that was already as old as your mother. Being born was just one step along the path of your development. Through gestation and childhood, every new experience changed you into something different than before. The building and rebuilding never stops, even as an adult. You replace your skin every four weeks, and make another 2 million brand-new blood cells every second.

Our age is something we take as a given, as clear as the calendar on the wall. No doubt it's a useful shorthand for our body's mileage and experience. But it's tricky to put an age on something that's dynamic and changeable. Our remarkable capacity for change continues throughout our lives.

2

GROWING UP

Rosemary wasn't buying it, and neither was I.

We were on a reconnaissance trip to the village of Ileret in northern Kenya, on the eastern shore of Lake Turkana, figuring out logistics to start a multiyear research project with the Daasanach community. Much of my research focuses on the intersection of lifestyle and health in farming and foraging populations, trying to understand how traditional subsistence communities like the Daasanach thrive in challenging landscapes. Despite the lake's proximity, the region is one of the hottest and driest on the planet. Sun-blasted sand and rock stretch into the distance in every direction, with scatterings of low, thorny bushes and parched trees afloat on an undulating sea of beige. Lake Turkana itself is a cruel tease, too salty and alkaline to make good drinking water. For an outsider like me it seems an unlikely place to call home, but pastoralists like the Daasanach have been tending to their herds of sturdy goats, cattle, donkeys, and camels here for thousands of years.

We had just been talking to the head of a German charity that had set up shop in Ileret years before to try and help the Daasanach. They had determined that malnutrition was rampant among the community and were working hard to make a difference. Children and their mothers were enrolled in an ambitious nutrition program, distributing high-protein, calorie-rich supplements to families in need. And they *all* seemed to be in need. The official numbers from the German charity, relayed in somber tones during our meeting, said that over two-thirds of the children were malnourished.

Rosemary and I sat back at camp on cheap plastic chairs beneath the dappled shade of an acacia tree, discussing the meeting. As a nurse, researcher, and outreach worker with the Kenya Medical Research Institute, she's been to every corner of Kenya and other parts of Africa, delivering health care to remote populations. She has seen her share of malnourished children, as have I. And it was hard to square the grim numbers from the German charity with what we saw with our own eyes.

We had spent the previous days traveling around Ileret, from the sandy dirt roads through the middle of the village to remote enclaves of small, windowless domed houses (the distinctive traditional Daasanach architecture) in the vast landscape beyond. Everywhere we went, children were running, playing, and laughing. Kids being kids. They didn't seem to be low on energy, nor did they seem particularly short, or "stunted." Sure, they were thin, but just about *everyone* here was thin, even the adults. The men and boys tending their herds, the women and girls fetching water from murky holes dug deep into dry riverbeds . . . they all had a characteristic build: tall and slim. It wasn't just the Daasanach. Most of the groups that call this region of Africa home seem to be built the same way.

Could it really be that nearly everyone in this region was malnourished? It was certainly possible. Life is hard for communities like the Daasanach, who feed themselves and their families from their herds. Hunger is never far away. Climate change has only made things harder, increasing the already brutal temperatures and messing with the seasonal rains. Many families probably weren't sure where and when they'd get their next meal, and the help from outside groups was no doubt appreciated. It was certainly well-intentioned.

But if malnourishment was as rampant as the German charity's numbers suggested, the signs should have been everywhere. Listless children. Bloated bellies. It should affect women's fertility, because starvation suppresses the reproductive system. None of that seemed to be the case. Energetic kids, young mothers, and big families were the norm.

Drawing with a stick in the sandy ground at our feet, Rosemary and I started plotting out a plan of attack. The German charity was tracking Daasanach kids' height and weight against standardized growth charts from the World Health Organization. That approach assumes all children follow—or at least, *should* follow—some uniform human pattern of growth. What if Daasanach kids grew differently? Maybe the skinny kids identified as malnourished were simply following their own path. Being slim could be a feature, not a bug.

Our research project had its first guiding question: How do Daasanach children grow?

MAKING YOUR BONES

Growth, whether you're Daasanach or Danish, is like any other kind of construction: Energy and raw materials from the food you eat are used to build more cells. New cells are layered on top of old cells, like brickwork. The more cells you build, the bigger you become.

All of our organs grow larger as we grow up, but it's the growth of our skeleton that determines our height. Your bones begin to form in the sixth and seventh weeks after conception. By twelve weeks, all the elements are there, the full skeleton in miniature. Apart from the skull and clavicles (collarbones), the elements are formed first in cartilage, which then slowly "ossifies," or turns to bone. The process is slow, spreading from ossification centers outward like ice growing on a pond. Many bones in the adult skeleton form from two or more ossification centers, growing and fusing together into a single element. The skull, for example, begins as five plates that eventually fuse together at their edges. In newborns those plates aren't fused yet, allowing the head to deform a bit during delivery as it squeezes through the mother's pelvis. A baby's skeleton is a mix of cartilage and bone.

By elementary school, children's skeletons are mostly bone, with the notable exception of the growth plates. The bones in your arms and legs, hands and feet are shaped like pipes with a cap on each end. Growth plates are the thin layers of cartilage separating those bony caps from the rest of the bone. Cells in the growth plates divide at a furious pace, continuously pushing the bony cap away. You can get the same effect by completely loosening (but not removing) the cap on a tube of toothpaste, holding it vertically with the cap end up, and giving it a squeeze: the paste growing out of the tube (representing the proliferating cells of the growth plate) pushes the cap upward, making the whole thing grow longer. The analogy isn't perfect—the cells of the growth plate don't come from inside the long bone but from the growth plate itself, and the toothpaste will quickly collapse under its own gooey weight. In a real bone, the growth plate cells ossify as they proliferate, maintaining just a thin layer of growing cartilage and keeping the bone stable.

One by one as we tumble through puberty, the growth plates fuse. The cartilage cells ossify faster than they can divide, and the whole plate turns to bone. There's a predictable pattern. Growth plates at the elbow fuse first, usually around twelve to fourteen years old; plates in the knees and ankles usually fuse around sixteen years; and so on. Forensic scientists and archaeologists use the pattern of fusion to deduce the age of adolescent skeletons. It's a bit of an educated guess, because, as anyone who has survived middle school knows, puberty and its many changes hit us at a range of ages. The growth plates of "late bloomers" will fuse at older ages, while the sixth graders who shave each morning will fuse earlier.

Once the growth plates fuse, a bone can't grow any longer. The cap has been screwed tightly onto the toothpaste tube and no amount of pressure can make it budge. The length of your bones, and with them your height, is set in stone. That's why a broken arm or leg in a child is particularly worrisome if the fracture occurs at the growth plate. Ossification at the fracture site can turn the growth plate into bone, preventing it from any further growth. Without careful intervention, the kid can be left with a child-length bone for the rest of his life.

Our bones also grow thicker as we develop, with bone cells called "osteoblasts" busily laying down new calcium-rich matrix on the surface of each bone like workers adding asphalt to an old road. Most bones are hollow inside, or filled with a honeycomb-like lattice of bone, which keeps the weight of the skeleton down while maintaining its strength. The hollows are excavated out by different bone cells called "osteoclasts," which eat away at the bone from the inside.

We can stimulate our osteoblasts to build more bone with

exercise and other physical activity. Osteoblasts can sense the forces the bone experiences, and they respond by building the bone stronger. In a classic example of this phenomenon, called Wolff's law after the guy who first noticed it, the racket arms of professional tennis players have thicker bones than their nondominant arms due to the repeated shock of thwacking a tennis ball. Archaeologists and forensic scientists use this same principle to reconstruct people's daily life from their skeletal remains. The dynamic nature of bone is one reason exercise is so important for kids—it helps grow the skeleton properly. Exercise can also help stave off bone loss and osteoporosis as we age.

The dynamic growth of bone affects the entire skeleton, not just our arms and legs. The jaw and the upper palate grow in response to the mechanical stresses of chewing, for example. This has led to a very recent problem: diets have been getting softer, leading to less forceful chewing and smaller jaws. The change from hunting and gathering (and chewing) wild foods to subsisting on softer farmed foods made smaller jaws and overbites the norm in the Neolithic period. Some have argued that smaller jaws changed the way we talk, adding "F" and "V" sounds to our languages. Today, with our mushy industrialized diets, our mouths often don't have enough room for our teeth, the sizes of which are under strict genetic control and not affected by diet. Our kids, growing up on milk-sopped cereal and creamy mac and cheese, end up needing braces and wisdom teeth extractions.

Your internal environment shapes your growth as well. Hormones are molecules made by the cells of one tissue that travel via the bloodstream to signal cells in another tissue—little messages in bottles floating in the ocean of our circulatory system, waiting to be found. For the message to connect, the receiving cell needs a specialized "receptor," a dock built specifically for that hormone. Without it, the hormone floats by, unread.

Growth hormone, as the name suggests, is a major player in determining how tall you'll grow. It's produced by cells in the pituitary gland, a pea-sized lump that hangs beneath the base of the brain like the knot of a balloon, nestled in its own bony pocket in the floor of the braincase. The region of brain above it, and connected to it by a stalk, is the hypothalamus. The hypothalamus is the overworked conductor of a complex symphony, coordinating nearly every facet of brain and body function behind the scenes. It works with the pituitary gland to produce *just the right amount* of dozens of hormones, including growth hormone.

Growth hormone is a good model for how many of our hormones are regulated. The hypothalamus produces a hormone called GHRH (growth hormone–releasing hormone), which then travels via blood vessels down the stalk to the pituitary gland. Cells there have receptors for GHRH, and are stimulated by it to produce growth hormone, which then circulates throughout the body in the bloodstream. In a genius bit of evolutionary engineering, when growth hormone circulates all the way back to the hypothalamus, it *stops* the hypothalamus from making GHRH, which *stops* the production of growth hormone. Only when growth hormone levels in the blood drop will the hypothalamus start making GHRH again. Systems like this are called "negative feedback loops," and we see them throughout the body.* It's a clever way to carefully regulate hormones, blood sugar, or other important substances so there's always enough but not too much.

Most cells have receptors for growth hormone, and as it circulates around the body it stimulates those cells to make proteins and divide. In kids, that leads to growth. Adults produce less growth

---

* The heating or cooling system in your home is another example of negative feedback: the thermostat sends a signal to the heater (or air conditioner), which stimulates it to produce hot (or cool) air. Once the air in the room meets the target temperature, the thermostat shuts off. It won't turn on again until the air temperature falls out of range.

hormone, but it helps cells repair the damage accumulated through daily wear and tear. A lot of this work happens overnight, which is an important reason children and adults need to get enough sleep.

The production of GHRH and growth hormone as well as their receptors is controlled by genes, highlighting just one of the hundreds of ways that genetic variation can lead to differences in height. Usually these effects are subtle, but occasionally they are stark. People with Laron syndrome, for example, typically don't grow much taller than four feet because the gene that builds their growth hormone receptors doesn't work. These individuals produce growth hormone just fine, but the cells throughout their body can't respond to it.

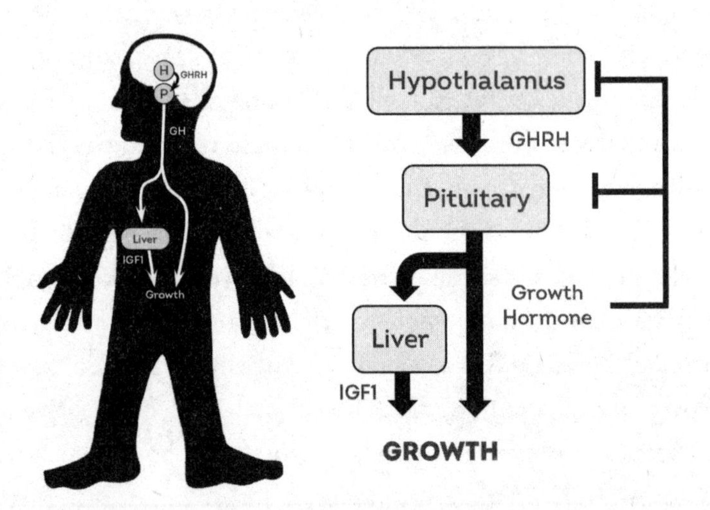

Figure 2.1 Growth hormone production and regulation. The hypothalamus (H) produces GHRH, which stimulates the pituitary gland (P) to produce growth hormone (GH). Growth hormone stimulates tissues to grow, and it also stimulates the liver to produce another potent growth-promoting hormone, IGF1. Growth hormone inhibits the hypothalamus and pituitary gland, keeping levels of the hormone in check.

Some kids don't produce enough growth hormone, falling off the growth charts as others their age shoot upward. This happened to none other than international soccer star Lionel Messi as a child. His parents, like millions of others in their situation, elected to give him daily growth hormone injections. It worked, sort of. Messi grew to five foot seven, below the average male height, but tall enough to be a monster on the soccer field. Other people produce too much growth hormone. Often the cause isn't genetic but rather a tumor or other insult that presses against or damages the pituitary gland. Many of the famous giants throughout history, including André the Giant, had this condition, known as acromegaly.

Growth hormone is far from the only hormone that affects how tall you grow. Achondroplasia, a more common condition than Laron syndrome that also causes dwarfism, occurs when the gene that makes the receptor for the hormone FGF isn't functional, affecting cartilage growth and leading to shortened limb bones. Thyroid hormone affects growth as well, by regulating cell activity throughout the body. And the sex hormones, testosterone and estrogen, shape growth, too. Testosterone promotes growth of both bone and muscle, and is the main reason men tend to grow taller and larger than women. (We'll discuss sex and gender differences in depth in chapter 7.) Hormone levels and growth respond to environmental factors like nutrition, but they are also influenced by our genes.

HEIGHT ADVANTAGE

Of all the qualities we look for in our elected leaders, height does not usually top the list. And yet, from George Washington (six feet tall), to Abraham Lincoln (six foot four), to Barack Obama (six one), U.S. presidents have all been taller than average, with very few exceptions.* We've had presidents across the political

---

* The diminutive James Madison, at five foot three, was the only president in over two hundred years who was probably *shorter* than average, though data on men's heights from his era are tough to come by.

spectrum, fat and thin, with varied backgrounds, from all over the country. The only consistent criteria for being leader of the free world seems to be the ability to reach items from a high shelf.

The benefits of being tall aren't limited to political office. On average, taller people get more education and report being happier. They tend to land jobs with higher status and better incomes. It's not just in the U.S. or other rich industrialized countries. Being taller is associated with better life outcomes in populations around the globe, including poorer countries. And it's not a recent phenomenon. Over a century ago, professor of commerce and finance Enoch Gowin, in his very serious and casually chauvinist guide *The Executive and His Control of Men*, wrote that men in charge tend to be tall. By his analysis, bishops were nearly two inches taller than small-town pastors, railroad managers an inch and a half taller than station agents, sales managers an inch taller than their salesmen. Governors were among the tallest, while musicians fared the worst. Of personal interest, I note that authors and lecturers compared poorly to more high-powered executives (for the record, I'm six foot one).

We seem to know all of this intuitively. We want our kids to grow tall, even injecting them with growth hormone to maximize their chances. And when it comes to our own height, we lie. The dating site OkCupid found that both men and women add an average of two inches to their real height on their dating profile. Everyone seems to do it, both shorter people and those already above average—even six-foot-three guys like me. But even with good nutrition and healthy hormone levels people end up at a wide range of adult heights. Much of that variation is due to differences in our DNA.

DNA is like a recipe book for making proteins, the fundamental building blocks of your body. Every part of you is made from proteins or assembled using proteins. You are human-shaped, and not ape-shaped, because the DNA in our lineage changed over the past 7 million years, resulting in a 2 percent difference between human and chimpanzee DNA. Any two humans on the planet today share over 99.9 percent of their DNA, but those tiny differences in DNA help explain why you look like you and I look like me. By dictating our proteins, DNA shapes our biology.

The DNA in your cells is packaged into forty-six chromosomes housed in the nucleus, plus a small loop in another part of the cell called the mitochondria. DNA is a threadlike molecule built like a long, twisted ladder (the famous "double helix"), and each rung of the ladder is made from two molecules called "nucleotides," or "bases," projecting from the sides and linking together in the middle (see figure 2.2). There are four nucleotides in DNA: adenine, thymine, cytosine, and guanine, abbreviated by the letters A, T, C, and G. A's always link to T's, C's always link to G's, so the sequence along one side of the ladder (say, GATTACA, like the 1997 sci-fi movie) tells us the sequence of the other side (in this case, CTAATGT), and we can discuss DNA as a single string of letters without worrying about the complementary strand on the other side. Your particular sequence of 3 billion A's, T's, C's, and G's is your unique, personal "genome"—the entirety of all the genetic information you contain.

To make a protein, the strands along a stretch of DNA separate, exposing the string of letters along one strand like the teeth of an open zipper. That sequence of letters acts as a template to make mRNA, a single-stranded chain of nucleotides similar to DNA. The mRNA, in turn, aligns a sequence of amino acids, which link together to form a protein. So, for example, the DNA sequence CGT will result in an mRNA sequence GCA, which will specify the amino acid alanine. Proteins, which are simply long chains of amino acids, can be useful on their own, or assembled into larger structures. Any stretch of DNA that makes a protein, or turns protein production on or off, is called a "gene."

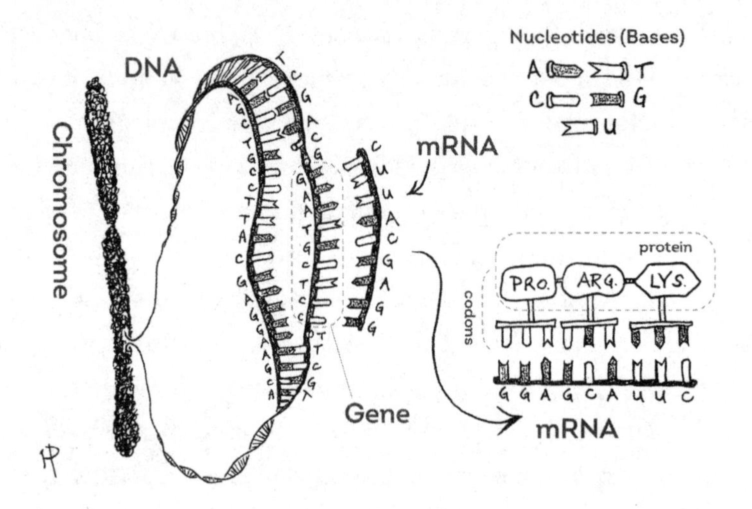

Figure 2.2 How genes make proteins. DNA is bundled into chromosomes. Its double helix structure unzips to expose a sequence of nucleotides (bases) that act as a template to assemble an mRNA sequence. In mRNA, the base tyrosine (T) is replaced by uracil (U). The mRNA then travels to the cell's protein assembly machinery, where it acts as a template for a sequence of codons, triplets of RNA that are each attached to an amino acid. The amino acids connect to form a chain, and that chain of amino acids is a protein. Sections of DNA that produce proteins are called genes. A typical gene is thousands of nucleotides long, and a typical protein is hundreds of amino acids long.

Height is the poster child for studying how genes shape our bodies: it's easy and fast to measure, and researchers have amassed datasets with millions of people, with each person's height and genome (their personal 3-billion-letter sequence of DNA). Height is a classic "complex trait," shaped by a combination of genes and environment, that follows a "normal" distribution: a few very tall people, a few very short, and a spectrum of folks in between. That wide range of variation and smooth, bell-curve distribution makes height amenable to statistical analysis. It's also noncontroversial, as traits go, even though the societal advantages of being taller are well known.

Height is strongly "heritable," meaning much of the variation in stature we see among people can be attributed to differences in their DNA. We all carry the same set of genes, but for every gene there are multiple variants, called "alleles," in the population. It's a bit like underwear: everyone has a pair, and they all have the same basic structure, but the styles might be different. Humans have twenty thousand protein-producing genes and thousands more regulatory genes, and each of us carries two copies of nearly every gene, because we get a full set from both our biological mom and biological dad. Your DNA—your genome—is unique because the particular combination of forty thousand–plus alleles you carry is unlike anyone else's (unless you've got an identical twin).

Heritability is a notoriously slippery concept in biology, but you can think about it as the accuracy with which we can predict a trait like height based on your alleles. A heritability of 0.0 means alleles don't tell us *anything* about why people differ in a given trait, while a heritability of 1.0 means alleles perfectly predict people's values for the trait. Heritability values for height are typically around 0.50 to 0.70 depending on the study, which fits with our daily experience. Tall parents tend to have tall kids because the alleles they pass down affect development in such a way that bones grow longer. Those alleles made the parents tall, and the copies passed down have the same effect in their children.

There are a couple of important things to keep in mind with heritability scores, which we'll encounter throughout this book. Low values don't mean genes aren't important, and high values don't mean environment doesn't matter. Heritability scores can't tell us anything about why two populations differ in a trait like height, because they are specific to a single population at a particular time and place. The heritability value for height would be lower in a population where access to nutritious food is uneven, because most of the variation in height would reflect children's access to food rather than their alleles. In a population where every kid has enough to eat, more of the variation in height would be due to genes and heritability would be higher. The difference in heritability values reflects the different circumstances, not the importance of genes or environment in shaping growth.

Height is also a textbook case in modern genetics because thousands of genes, and the alleles you carry for each one, have an impact. Over the past two decades, we've come to understand that most traits are influenced by the alleles we carry at dozens, hundreds, or even thousands of genes, each with tiny effects. Identifying all those alleles required a major development in modern genetics, the genome-wide association study, or GWAS. GWAS compares the genomes of hundreds of thousands, or even millions of people, to search for single-letter variants, like having a T instead of a C at a particular location, that are associated with a particular trait. The underlying logic is that each variant (a particular kind of allele called a single nucleotide polymorphism, or SNP, pronounced "snip") is a reliable indicator of a gene allele carried at that location. The SNPs themselves aren't necessarily functional, but instead act like Post-it notes stuck in the pages of an immense book, flagging the important passages.

The most comprehensive GWAS study of height to date identified over twelve thousand SNPs associated with height. SNPs can overestimate the number of genes at play for a given trait because it's possible that two or more SNPs could flag the same gene. But even with that caveat, it's clear from GWAS studies that there are thousands of genes working together to shape how tall you grow.

The large number of genes associated with height tells us two things. First, most of the genes that affect height have small, nearly imperceptible effects. You might recall genetics lessons from high school biology, in which the color of a pea plant is determined by the alleles for a single gene. As we've come to learn over the past couple of decades, simple traits like those are vanishingly rare. For the typical gene that affects height, having one allele versus another will impact your adult stature by less than one millimeter. Only when you add up those effects over the thousands of genes do you have any sense of whether you're likely to be tall or short.

The second thing that the huge number of height-related genes tells us is that *a lot* of processes are involved in determining your height. Pretty much everything a cell does, added up over the 37 trillion cells in your body at work every day, can affect growth. It turns out that many of the traits we're most interested in, like disease risk, propensity for obesity, athletic ability, or intelligence, are similarly "polygenic," influenced by hundreds or thousands of genes, each with miniscule effects on multiple aspects of our biology.

Genes that affect growth don't just influence size, they can also determine shape. Genes affecting cell division, hormone sensitivity, and the like can cause different regions to grow thicker or thinner, more pronounced or fine. That's why your nose looks like your mom's, or you have your dad's chin dimple. More than two hundred genes have been identified that contribute to facial shape, for example. These genes are why people from different parts of the world have distinctive facial features. My wife's Asian heritage is as obvious from her face as my enormous European nose is on me, and those differences are genetic. Our kids, with a fifty-fifty mix of our alleles, have features that fall somewhere in between.

I've heard lots of stories over the years as to why certain features are common in different parts of the world, running the gamut from good-natured folk stories to evidence-free scientific conjecture to thinly veiled racism. There's a long and often laughable history of scientific explanations for facial variation going back to Darwin. Some have argued, for example, that the "epicanthic fold" that gives Asian eyes their distinctive shape is an adaptation to cold climates in high northern latitudes, but it's not clear that structure has any meaningful effect on thermoregulation, and that same feature is found in many warm-climate African populations as well. The reality is that most of the global diversity in face and head shape is probably genetic noise, the random slosh of alleles as populations spread and mixed across the globe.

This brings us to our first lesson about human diversity. Most of the variation we see among people and populations is random, with no underlying evolutionary cause or meaningful impact on the way the body functions. New DNA mutations pop up with each generation, creating new alleles that get circulated around the globe through migration and intermarriage—what biologists call gene flow. This sort of mixing inevitably leads to pockets where some alleles are more common than others, like clumps of pretzels or peanuts in a bowl of snack mix. Geographic patterns in alleles or traits can be helpful for guessing a person's ancestry (that's how companies like 23andMe do it), but they don't necessarily impact the way our bodies function in any meaningful way.

Which leads us to the second lesson about our diversity. Occasionally, variation in a trait *does* reflect some functional effect that was favored (or disfavored) in a particular environment. Noses seem to be an example. Alleles that build large noses like mine appear to have been favored in colder environments, perhaps because they help warm and moisturize cold dry air before it hits the lungs. An alternative explanation, and one that's hard to test, is that large noses are a product of "sexual selection," in which a feature becomes widespread because potential mates find it attractive, like a peacock's tail. My gut tells me it's probably climate, but

I like to imagine that a large nose was, at least in some population, long ago, absolutely irresistible.

Climate also seems to have had an effect on our body proportions. Populations near the equator tend to be taller and more slender, while Indigenous groups living in colder climates closer to the poles tend to be heavier, with shorter arms and legs. We see this pattern, known in biology as Allen's and Bergmann's rules after the nineteenth-century scientists who discovered it, in other species with wide geographic ranges as well, and it's thought to be an adaptation to maintain a constant body temperature. In hot climates, a tall, thin build increases surface area and makes it easier to radiate excess heat. In cold climates, a larger body generates more heat and shorter limbs help keep it in.

Nose shape and body proportion are examples of local adaptation in humans, in which the alleles for a particular trait (a slim build, a big nose) are favored by evolution because they improve a person's ability to survive and reproduce, passing those alleles to the next generation. Over time, those favored alleles become very common in a population. This is Darwin's big idea, "natural selection," at work. If the trait involved is visible, like nose shape or stature, local adaptation can give populations a distinctive look.

But we need to resist the temptation to attribute every feature, every difference between populations, to local adaptation through natural selection. Usually, the folk wisdom around biological diversity is just storytelling. Most variation we see within and among populations is just random noise, with no impact on survival or reproduction.

So how do we separate random, neutral variation (which is incredibly common) from local adaptations (which are much, much rarer) when we look across the full diversity of our species? There are sophisticated mathematical tests that biologists can use to test for the telltale signs of adaptation in a gene or a trait. But helpfully, there's an easier test, nearly as effective, that we can apply to any trait. These three criteria must be met for a local adaptation to evolve through natural selection:

1. The trait must be heritable, with at least some contribution of genetics to the way the trait develops. If the variation in the trait is purely due to environmental factors (for example, whether you speak English or Russian as your native language), then it can't evolve through natural selection.
2. The trait must have a meaningful impact on reproduction. Survival is important, too, but only insofar as it improves your chances to reproduce. That's how the alleles that shape the trait in a favorable way (say, by making you tall and thin in a hot climate) become more common in subsequent generations: people carrying those alleles will have more successful offspring who also carry the favorable alleles.

Criteria 1 and 2 are necessary for any trait to evolve by natural selection. They are Darwin's evolutionary engine, the force that gave giraffes their long necks and cheetahs their speed. For a local adaptation to evolve, a trait that differs among populations within a species because it's adapted to different environments, one additional test must be met:

3. The evolutionary pressure shaping the trait must be localized and consistent, with stable differences between regions that last for many generations, typically for thousands of years. If local conditions aren't meaningfully different, or those differences don't persist long enough for natural selection to change the alleles in a local gene pool,

then intermarriage and migration (gene flow) will erase any population differences.

Body proportions meet all these criteria because the trait is heritable, maintaining body temperature is essential to survival, and the evolutionary pressures at work (hot environments near the equator, cold climates near the poles) have been consistent for millennia and differ geographically. We'll need to keep these criteria in mind as we investigate other biological diversity throughout this book.

LIVING UP TO YOUR POTENTIAL

Geography doesn't just affect the alleles for body proportions. Where you live can have big effects on how tall you grow. Classic studies of migration have demonstrated repeatedly that children who leave poor economic conditions for more affluent countries typically grow taller than those who remain in their homeland.

Environmental effects on growth generally come down to energy: if the body doesn't get enough calories to support growth, or is forced to spend too much of its energy on strenuous work or fighting disease, it won't have the energy to grow. In nineteenth-century England, it was well-documented that kids sent to work in factories were about four inches (ten centimeters) shorter, on average, than their upper-class peers. Not only were those kids working long, hard hours each day with little food, they were often housed in unsanitary conditions that promoted disease. They weren't getting enough to eat, and the calories they did manage to get were going to factory work and their immune systems. They didn't have the energy to grow.

In the past, prior to the miracle of modern plumbing and the widespread use of soap, the challenge of getting enough calories to grow was common. The transition to farming seems to have been particularly difficult. Farming generally produces more food for less effort than hunting and gathering, which is one of the reasons agriculture replaced foraging as the dominant way of life around ten thousand years ago. Those extra calories meant more energy for reproduction, and a Neolithic baby boom followed. Skeletal collections from archaeological sites in Europe, Asia, and North America show big increases in the numbers of children being born with the adoption of farming.

You might expect that all those extra calories would mean better growth as well, but that's not what we see. With more people living in densely settled farming villages rather than smaller, more mobile hunter-gatherer camps, infectious disease spread much more easily. Fertility increased, but so did childhood mortality and disease. Kids of that era likely spent the majority of their childhood fighting off viruses, bacteria, and parasites. All that energy spent on immune function meant less was available for growth. From measures of the skeletons at archaeological sites before and during the transition to farming, we know that average adult height declined substantially, by three or more inches in some populations.

Small-scale farming communities today provide even more insight into the battle between immune function and growth. The Shuar population in Ecuador and Peru numbers about forty thousand, with many living in towns and small cities, while others live in tiny, remote villages. Those in the remote villages make a living farming plantain and other crops by hand, as well as hunting, fishing, and foraging wild foods. People in those villages, including the kids, deal with all sorts of pathogens, from bacteria and viruses to worms and other parasites. Work led by anthropologist Sam Urlacher, now at Baylor University, connected the dots between all this immune activity and childhood growth. Using finger-prick blood samples to track pathogen burden, along with careful weekly measurements of leg length (which picks up activity at the grow plates in the leg), Sam showed that Shuar kids fighting infections grew less. And these were not kids who had a fever or were obviously ill. Most of this happened under the radar, the immune system silently at work.

One bright spot in global development has been the improvement in basic nutrition and hygiene across most of the world. Kids are growing taller. Food shortage and famines remain a real danger for far too many, but in the large majority of countries, all across the planet, boys and girls are growing an inch (2 to 3 cm) taller today than in the 1980s.

In fact, in the developed world, for families with regular access to nutritious food and effective medicines (including antibiotics and vaccines), stature is approaching people's "genetic potential," the maximum expected for their particular assortment of height-related alleles. Adding more calories won't improve stature any further, as growth is already maxed out. We see evidence for this in countries like the Netherlands. The Dutch are the tallest population in the world, and their average stature has been increasing steadily over the past century as living standards improved. But recently, this growth has plateaued. The same is happening in other developed countries all over the globe. We're hitting our heads against our genetic ceiling.

As we touched on in the last chapter, childhood nutrition does more than promote growth. Kids with enough to eat are less likely to develop heart disease, diabetes, some cancers, and a range of psychological and cognitive problems as adults. There's reason for optimism that some of these problems might become less common as today's well-fed children age into adulthood and their golden years. It's a tricky balance, though. The major nutritional challenge facing the developed world isn't ensuring that our kids eat enough, it's making sure they don't eat too much. In fact the same study that found heights increasing globally since the 1980s found that weight has been increasing at an even faster clip, with a growing proportion of children overweight and obese. We'll tackle the obesity pandemic in chapter 5, but it's important to recognize the crucial role of childhood nutrition.

WINNING THE GROWTH LOTTERY

Can the biological control of growth explain the *benefits* of being taller?

The standard story about why taller people reap so many societal benefits is that our evolved brains still hold a preference for bigger people. We evolved as hunter-gatherers, sharing and cooperating, relying on one another for food and protection. It stands to reason that bigger people would have made better partners. A preference for taller people could have evolved, as the tendency to seek out taller people improved a person's chances of survival and reproduction, the currency of natural selection. We are also uniquely cultural animals. Favoring taller people could be learned, a relic of Paleolithic culture that remains common today.

Whether it's learned or genetic, there are a couple of ways our preference for taller people could play out. They could be treated better throughout their lives—judged, unconsciously no doubt, as better or smarter by their teachers, picked for leadership roles early in their careers, favored by their bosses for the raise or promotion. In this scenario, taller people benefit from a rigged system, through no fault of their own. The "height premium," as economists call it, is just dumb luck and our out-of-date, Paleolithic preference for larger partners. An extension of this explanation is that taller people could develop better self-esteem from a lifetime of others' favor. They internalize all the positive feedback others give them, and they believe they really are better. They tend to get the promotions, the raises, and the romantic partners because they have the self-confidence to go after them.

We see this with facial attractiveness, by the way—another example of growth patterns affecting our lives. As we discussed above, the shape of your face is determined largely by your genes. The unique assortment of alleles you inherited from mom and dad affect the size of your cheekbones, the length of your nose, the shape of your eyes. There's also the element of chance in how symmetrical the right and left sides of your face become as you develop from an embryo. And of course there's the cosmic roll of the dice as to whether your particular set of features will be deemed attractive by society. If they are, you're in luck. Unless you are both gorgeous and clueless, it will not surprise you to learn that there's a "beauty premium," much like that for height. Beautiful people get promoted, earn more money, and generally report a better quality of life. In careful lab experiments where people negotiate for money, we let them take an unfairly large share.

A more provocative explanation for the height premium is that taller people are *actually better* than short people, at least on average. The argument is that tall people had all the nutritional support they needed to reach their full genetic potential growing up, while shorter people didn't. Height, in this scenario, would be what biologists call an "honest signal" of ability. It seems far-fetched, but given what we know about the impact of prenatal and childhood nutrition on development, including cognitive ability and mental health, it's not completely out of the question. Some studies show taller people tend to score higher on IQ tests, and there is a correlation between years of education and stature. But we also know that subtle differences in the way kids are treated can have effects on IQ scores and educational results from an early age, something we'll dive into in the next chapter. I'm not convinced we can easily disentangle the way society treats taller people (including kids) from any real, biological advantage they might develop.

Another problem with the "honest signal" argument is that most kids in developed countries get plenty to eat. If that's the case, differences in height are more likely to reflect our genetics than our nutritional environments growing up. The fact that we see a correlation between height and IQ scores or educational attainment today, among wealthy populations, makes it less likely that those relationships reflect meaningful differences in childhood nutrition and more likely we're seeing the societal preference for height play out in scholastic ability. The beauty premium would seem to support this view. No one would argue that your looks determine anything meaningful about your abilities.

The way human societies around the world treat tall and attractive people differently can feel a bit depressing, particularly if you're not thrilled about the features you've been dealt. Your height and looks are largely beyond your control, the product of your DNA and the environment in which you grew up, and there's a large element of pure chance in both. (Which alleles did you get from mom and dad? Were you lucky enough to avoid serious illness and be born into a family with plenty to eat?) If you're an adult, and your growth plates have fused, there's not much you can do to change your height aside from wearing heels. There's more latitude for changing your looks (the $530 billion cosmetic industry is here for you), but let's face it, there are limits there, too. As my dear old grandmother liked to say, you can't make a silk purse out of a pig's ear.

But cheer up! Height and looks are hardly your destiny. As we'll see in the next chapter, when we discuss the genetics of IQ, a statistical correlation does not mean that we can predict how your life will turn out from your height or your looks. Jeff Bezos, billionaire owner of Amazon, is only five foot seven (170 cm), as was Albert Einstein. And while both were attractive enough to find spouses and reproduce, neither is widely considered a sex symbol. There's hope for us all.

BACK IN KENYA

Rosemary and I mulled over the puzzle of Daasanach growth for the remainder of the trip, thinking about what we'd need to solve it. To determine if malnourishment was as widespread as the German charity believed, we'd need to build our own growth charts for Daasanach kids, and that would take time. Even in a relatively small population, where the kids grow up in similar conditions across households, there's enough variability from child to child that it takes hundreds, even thousands, of participants to develop robust growth charts. It was going to take years.

But sometimes luck finds you. For nearly a decade, the head nurse and community health volunteers who staffed the small village clinic had been collecting heights and weights of area children as part of their vaccination and nutrition programs. We noticed the dusty record books, stacked on sagging shelves, during one of our first visits to the village health clinic, a simple cement building that was showing its age. Curious, we cracked one open and were instantly agog. Each one was filled with thousands of entries, children's heights, weights, and health status printed neatly on row after row in careful blue ballpoint pen.

Could we use the records to build community growth charts? we asked.

Sure, that wouldn't be a problem, said Beatrice, the nurse in charge of the clinic. No one else had ever paid the records much attention. She seemed happy that they might finally be of use.

We couldn't believe it. Thousands of hours of carefully

collected data had just fallen in our lap. Zane Swanson, a PhD student in my research group, led the charge, digitizing the handwritten records and conducting additional fieldwork to develop a broader understanding of health and lifestyle. When the dust finally settled, we had a solution to the puzzle of Daasanach growth.

Daasanach children are born the same weight and length as babies all over the world. Compared to World Health Organization international growth charts, the Daasanach are right where we'd expect any newborns to be. But their pattern of growth soon diverges. For the first year, they slowly but steadily fall behind their peers in other countries for height and weight. They aren't growing as fast. This, in itself, wouldn't be cause for too much alarm—the difference isn't huge, and we see similar patterns in some other communities. Importantly, the kids' proportions, their weight-for-height, is steady. They don't look too thin.

Then, at about age two, something remarkable happens. Their height takes off, shooting past their international peers. The growth plates in the Daasanach children's limbs are manic, adding bone and increasing their height at a pace we don't see in other groups. By age five, Daasanach boys and girls are taller, on average, than children from other countries, even those with easier access to food. Their weight increases as well, but not at the same impressive pace. The classic Daasanach body proportions, lean and tall, are taking shape.

This was Allen's and Bergmann's rules in action, and a clear example of local adaptation. The alleles for tall, slim builds, favored over thousands of years in hot climates to prevent overheating, were at work behind the scenes, manipulating hormones and receptors. Alleles that helped their ancestors thrive in a tough environment were the children's genetic inheritance. Daasanach children weren't falling behind, they were following instructions.

But that's not the lesson the German charity had taken. By

comparing Daasanach children to international standards, they had assumed that the alleles for growth and proportions were the same for everyone, and that any deviation from that international standard was due to the nutritional environment. In effect, they were ignoring the possibility of local adaptation. They were alarmed by the children's weight-for-height ratio, which was well below kids in Europe or Asia. But it wasn't due to a failure to grow. Daasanach kids were growing so tall, so fast, that the ratio of weight to height was *always* going to look low. You could pour as many food supplement calories into the population as you wanted to and you wouldn't change those cleverly adapted body proportions. The German charity was destined to fail because they misunderstood how humans adapt.

Race-based views of human diversity would have missed the mark as well. Local adaptations, when they occur, affect regional populations. The broad categories we typically use in the U.S. for race and ethnicity, like Black, White, Asian, and Hispanic, do not refer to coherent, regional populations. They are culturally constructed categories that each encompass an enormous range of biological diversity. Migration and intermarriage over the past twenty thousand years have resulted in complete genetic overlap of these groups, with no meaningful way to define them biologically. Even a seemingly remote population like the Daasanach shares over 90 percent of its alleles with other populations around the world.

To see just how wrongheaded a race-based view can be, look no further than Rosemary, me, and Beatrice, the local woman who runs the Daasanach community health clinic. In the U.S., Rosemary and Beatrice would both be considered Black, with the implicit assumption that they were somehow more biologically similar to each other than either was to me, a White person. In fact, there is so much genetic diversity within Africa, and there has been so much global movement in the past, that it's just as likely that Rosemary and I share more alleles than either of us share with Beatrice, or that Beatrice and I share more alleles than either of us share with Rosemary. Even if Beatrice and Rosemary do share more alleles, racial categories would be worse than useless for making sense of our biological diversity. In some traits, like skin color, Rosemary and Beatrice are clearly more alike, but in others, like body proportions, Rosemary and I are more alike. In still others, I might be more similar to Beatrice. Race tells us nothing about these similarities, because race isn't a meaningful biological category.

The real shame of the German charity's approach wasn't that they wanted to help, it's that they didn't really know how. They didn't understand diversity and were ignoring the possibility of local adaptation for the traits they were most interested in, height and weight. With community-specific growth standards, we can do a better job identifying the kids that are truly falling behind and work to get them the support they need. Understanding how genes and environment affect the way our bodies grow and function doesn't just help us see ourselves, it can help us lift up others.

At least the charity's efforts were well-intended and didn't cause any obvious harm. The food supplements they handed out were no doubt appreciated, and there were surely many children who benefited from the extra nutrition. Other societal programs for improving public well-being haven't always been so kind.

3

HEAD QUARTERS

This was bad, worse than expected. Over halfway through my second IQ test, and I was hopelessly adrift. As a scientist, someone who makes a living by figuring things out, it's a point of pride that I'm quick on my feet, that I can solve puzzles, spot patterns, and come up with clever solutions. Looking at the exam in front of me, I had no clue where to even begin. It had been this way for hours. A darkness began to seep into my thoughts . . . *Maybe I'm just stupid.*

The nice man giving the exam seemed to have reached that conclusion much earlier. I noticed he had stopped pointing out obvious mistakes or repeating the questions. He was friendly, even a bit chatty when we started, but now I was lucky to earn a thin smile.

In my defense, this wasn't the usual paper-and-pencil test I grew up with. We were in northern Tanzania, and the test was being administered by a man from the Hadza community. The Hadza are traditional hunter-gatherers, clever and resilient folks who make a living from the wild plants and game around them. They live in small camps of simple grass houses on the dry, open savanna of northern Tanzania. Hunting and gathering is hard work, both physically and mentally. Every day brings a new challenge, a test to pass, and the scoring is simple: solve the puzzle and you get to eat.

Today's exam was tracking a wounded impala, shot earlier that afternoon with a poison-tipped arrow through its midsection. As usual, the impala had bolted when it was shot, and was out of sight and far away from the hunter within seconds. Rather than push it farther by giving chase immediately, the hunter, Dofu, knew to let it run, coming back to camp to gather his buddies to track it down later. Now, combining information from hoofprints, blood splatter, and other signs of the impala's passing with what we knew about impala behavior and the tendencies of a gravely injured mammal, we had to locate the carcass. We had left camp at a brisk walk, intercepting the trail about half an hour out. A series of questions followed, posed by the tracks and other signs: Which way did the impala go? Was it laboring, close to death? Where was it trying to get to?

Dofu and I were joined by five other men, friends who enjoyed a good puzzle and the taste of fresh impala. We—or actually, *they*—would find a stray hoofprint in a patch of soft soil, a drop of blood on a rock, a broken blade of grass. *How long ago was the animal here? Where was it headed?* I would try to concoct an answer, an inkling, a reasonable guess. Wrong every time. The men, with barely a moment's digestion of the evidence, would be off again in unanimous agreement, inevitably in a different direction than I would have chosen. Soon we'd find more signs, proof that they had answered correctly, while I had not. On a typical multiple-choice test you're bound to get some right just out of pure dumb luck. I'm not sure I was achieving even that level of incompetence.

The standard excuse for bombing an IQ test is being unfamiliar with the material. Sadly, that wasn't the case. I grew up hunting deer in rural Pennsylvania, a pretty similar task to hunting impala. I had spent plenty of hours following tracks and blood trails with my dad, uncles, and cousins. And I had spent time with the Hadza before, and had tracked impala and other game. I had even completed college- and graduate-level coursework in animal ecology and behavior.

Yet here I was, failing on the two dimensions of intelligence that most IQ tests aim to measure. My "crystallized intelligence," the deep well of stored knowledge about the subject at hand, was an embarrassment. I had no idea what to expect from a wounded impala, how to read the landscape around me from its perspective, or how far or fast it might run. My "fluid intelligence," using patterns and general principles to make predictions and solve puzzles, was equally terrible. I looked at the same tracks, the same dried blood, the same rocky hillside as the other guys, and where they saw a complete and detailed story I drew a complete blank.

My first IQ test had gone much better. In first grade, as a peculiar six-year-old, I had sat with a friendly examiner for an hour or so, answering questions and solving puzzles. I remember using black and white shapes to re-create the patterns he showed me, and telling him what was missing or strange about a picture or a story. I had done well enough to be placed in the "enrichment" program, which I loved because it made me feel smart and got me out of normal, boring school on Wednesday afternoons. I suspect it helped to bend my trajectory toward a career in science.

Here with the Hadza, things were decidedly different. It wasn't just tracking that found me intellectually deficient. Men and women both had an encyclopedic knowledge of the landscape around them, the names and uses of all the plants and the behavioral and physical traits of all the animals. They could work out how to build and repair everything they needed, from rainproof houses to sturdy bows and arrows, even things they'd never seen before (I once watched a Hadza man successfully rewire a broken FM radio with tools he made himself). I would go out foraging with the women and stare in wonder as they pounded the ground with rocks to determine the size and ripeness of the starchy tubers growing beneath the surface. To this day I don't have the foggiest idea what they were doing—were they listening? or feeling? both? neither? I felt like a dog watching NASA engineers plotting orbital trajectories.

WE COME INTO THE WORLD WITH OUR BRAINS UNFINISHED. AS A RESULT, we're completely helpless. A newborn gazelle hits the ground and is off and running in minutes. We can't even hold our own head up for months. It's years before we can walk, talk, and reliably control our bladders. As hard as it might be for new parents to believe, those helpless years are a feature, not a bug.

The earliest members of the human lineage were essentially bipedal apes, creatures with a chimpanzee-sized body and brain that moved through the world on two feet, but lived a generally apelike existence. Then, about two and a half million years ago, things began to change. Populations in East Africa started hunting and scavenging more, incorporating more meat into their diet. This was the beginning of hunting and gathering and of our genus, *Homo*. Hunting and gathering was a game-changing strategy, the first in the history of life: one species with a two-pronged foraging approach targeting complementary foods, cooperating, and sharing.

It was a strategy that could succeed anywhere, and Paleolithic *Homo* populations became so ecologically flexible that they could thrive in nearly any habitat. By 2 million years ago, the species *Homo erectus* had spread across Africa and into Eurasia, turning up at archaeological sites as far east as Indonesia. Novel environments and a reliance on their wits put an evolutionary premium on intelligence and creativity. Material culture developed alongside increasingly larger brains.

By the time our species, *Homo sapiens*, emerged in Africa roughly three hundred thousand years ago, each new generation was inheriting much more than just their parents' genes. A rich and complex cultural heritage was theirs as well, shaping every aspect of life and deeply intertwined with their biology. Techniques to hunt and gather. Methods for making fire, tools, and clothes. Social rules that governed daily interactions, from friendships to marriage. Culture determined what we ate, who our allies were, and how we reproduced. We live our lives atop a vast mountain of accumulated cultural knowledge.

The sheer volume of information we each need to learn changed the way our brains develop. Rather than entering the world with a brain full of hardwired instincts and ready-made strategies, we arrive unfinished and built to learn. Our undercooked brains are sponges. There is so much to learn, no matter where you grow up, that the process takes a long time. Hadza kids can't hunt and gather proficiently until their late teens. I couldn't pay my own rent until I was in my twenties. But while our long period of dependency and learning is universal, the results could not be more different. Some of us learn to track impala, while others (not naming any names) do not. Within our own community, some flourish in school, while others struggle. Some are drawn to working with their hands, while others prefer abstractions. Within one family, a boy might write symphonies at ten years old, while his sister prefers to spend her free time playing soccer.

Our brains define us. Your personality, preferences, intelligence, temperament, and talent all emerge from the cacophony of cells inside your head. The way we understand our brains shapes the way we understand ourselves and our differences.

WIRING DIAGRAMS

The brain is a pale gray three-pound (1.3 kg) blob with a consistency somewhere between wet scrambled eggs and cheesecake. Slice it in half and you'll find the gray color is like a rind, with most of the interior of the brain a ghostly white. Both gray and white areas look deceptively homogenous and soft, without any clear inner structure. The gray area is actually an incredibly dense network of microscopically thin threadlike cells called neurons. Your brain has 86 billion neurons, all woven together. If you could stretch them out end to end, you'd have around 300,000 miles (500,000 kilometers) of brain wiring, enough to reach past the moon. The neurons are embedded in a matrix of fatty support cells called glia, the white stuff, which outnumber the neurons nearly ten to one. The glia don't do any thinking, they're only there to help with housekeeping tasks like delivering nutrients to the neurons and guarding against infection. It's also glia that police the "blood-brain barrier" that filters which molecules are permitted to reach the neurons. Your neurons are like a dense tangle of Christmas lights set in a Jell-O casserole of glia.

Just like string lights, neurons send electrical signals to one another. Instead of a long continuous wire with lots of little bulbs projecting from it, imagine the string of lights as a chain of individual units that begin at one bulb and then end at the next. Each light bulb in our Christmas-light-brain-cell metaphor is the body of a separate neuron, and the wire projecting from it is that neuron's "axon." The body (or bulb in our analogy) houses the nucleus and other important bits of cellular machinery, while the axon is just a long, thin wire. When the body is stimulated by a signal coming in from the neuron before it (which we can picture as the bulb lighting up), it sends an electrical signal down the axon to the next bulb. If it sends a strong enough impulse, the next bulb will light up and send the message onward, down its axon. And on it will go, the electrical signal sent from one bulb to the next down the chain.

The electrical signal sent along the axon is a cool bit of evolved physiology called an "action potential" that we find in all animals and in a couple different cell types, namely neurons and muscle cells. The axon is essentially a long, thin tube, like a garden hose. There's fluid (mostly water) both inside and outside the axon, but there's an important difference on either side of the axon wall. Inside, the fluid is filled with a lot of charged potassium molecules. Outside, the fluid has a high concentration of charged sodium molecules. Because of the sodium-potassium imbalance, the inside of the tube is negatively charged, with a higher concentration of electrons compared to the outside.

This setup is like a mousetrap ready to spring: the laws of physics want to balance out both the charge difference and the sodium-potassium difference on either side of the axon's wall. All that's needed is for something to open a door to let those charged sodium and potassium molecules race past each other to balance everything out.

When the neuron body is stimulated (the light bulb lights up), this is exactly what happens. The stimulus adds a little positive charge to the fluid inside the axon, right at its base where it leaves the neuron body. If the stimulus is strong enough, that bump in charge will cause channels on the axon wall to open, allowing sodium to rush in and potassium to rush out. For a moment, the fluid inside the axon becomes positively charged. That causes channels in the next section of axon, a bit farther down from the neuron body, to open. Sodium and potassium rush past each other there, causing channels in the next section to open. . . . And on it goes in a chain reaction down the length of the axon, a wave of positive charge and open channels rippling toward the next neuron. That's an action potential, and they can reach speeds in excess of 250 miles per hour (120 meters per second).

The movement of charged particles across the axon wall makes an action potential an electrical impulse. That's why we can monitor brain activity by measuring electrical signals, and why external electrical impulses (say, a shock from a frayed extension cord) can mimic action potentials, for example causing muscles to contract. It's also the basis of electroconvulsive therapy, previously called electroshock therapy: electrical current is sent into the brain with the hope of disrupting its circuitry and resetting it to something better. And in epilepsy, seizures are caused by electrical storms that erupt within the neural circuitry and reverberate out to the rest of the brain.

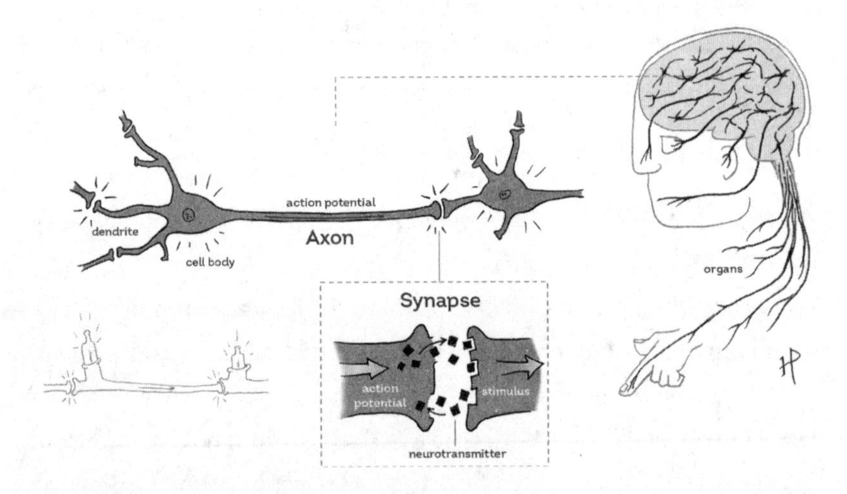

Figure 3.1 Neurons communicate like string lights. A stimulus is received from another neuron (or a special sensory cell, like those in the eye) at a dendrite. If it's strong enough, the stimulus will start an action potential that zips down the axon toward a target neuron. At the connection point, called a synapse, neurotransmitter molecules flood out from the sending neuron, are sensed by the target neuron, and are then reabsorbed.

The end of the axon connects to the next neuron in the chain at a junction called a "synapse." When the action potential reaches the synapse, the electrical stimulus causes the tip of the axon to puke out a bunch of molecules called "neurotransmitters" into the junction between the neurons, where the receiving neuron sucks them in and becomes stimulated. If the stimulation is big enough, that neuron will send its own action potential down its axon, and the process repeats itself.

There's a long list of neurotransmitter molecules, and different neurons specialize in particular types. You've likely heard of some of them, like serotonin, dopamine, and oxytocin. Brain-altering drugs, both pharmaceutical and recreational, mimic or manipulate these neurotransmitters to achieve their effects. Opioids like heroin mimic molecules your body makes naturally, binding to neurons to inhibit their ability to produce an action potential, calming everything down. Nicotine from tobacco binds to neurons and increases the production of dopamine, which generally makes people feel good. Selective serotonin reuptake inhibitors, like Prozac or Zoloft, cause serotonin to linger a bit longer in the synapse before it's vacuumed back up by the axon (axons recycle their neurotransmitters), prolonging the positive feelings associated with serotonin signaling.

The problem, of course, is that these drugs can become addictive. Neurons are dynamic, changing constantly in response to new conditions. That flexibility allows us to learn and adapt, but it means our neurons quickly change to anticipate the new normal. In people who smoke, vape, or chew tobacco, their neurons respond to the regular flood of nicotine by becoming less sensitive to it. As a result, people addicted to tobacco *need* to flood their brain with nicotine to produce a normal amount of dopamine—without it, they feel anxious and irritable. And they learn very quickly to associate tobacco use with dopamine release and crave a cigarette. Other addictions often work the same way, with the brain adapting to the new dopamine or serotonin trigger—a social media hit, a blackjack table, or a drug—and rewiring itself to expect and rely on them.

LEARNING, EVOLVING, AND THE HARD PROBLEM

Our string lights and cigarettes model of neuron networks gets us surprisingly close to a real understanding of how the brain and the rest of the nervous system work. Nerves, which are bundles of neurons, project out from the brain or spinal cord and into every nook and cranny of our bodies, millions of tendrils reaching everywhere. Those neurons are constantly bombarded with stimuli from both inside and outside the body. There's the five senses we're familiar with, sensitive to light, sound, touch, and the molecules we identify as flavors and smells. But there's dozens of others at work throughout the body. Neurons embedded into blood vessels, sensitive to blood pressure and the amount of carbon dioxide in the blood. Neurons in the gut, sensitive to the stretch of a big meal or the arrival of sugar and fat. Neurons in the skin, sensitive to temperature. The list goes on.

Each of those millions of neurons send action potentials along their axons to the brain (neurons that live below the neck send signals to neurons in the spinal cord, which then relay them to the brain). In the brain, those signals are routed to particular regions dedicated to specific tasks. Most of this processing happens below the level of conscious thought. When neurons embedded in the blood vessels sense high levels of carbon dioxide, that action potential signal is routed to the brain stem, near the base of the spinal cord. Neurons there respond by sending an action potential signal back down into the body to the diaphragm, the main muscle involved in breathing, to stimulate breathing and clear excess carbon dioxide from the blood.

We can see how these simple stimulus-response circuits can

shape behaviors. You are born with a handful of prewired preferences, neural circuits that connect a particular stimulus to the release of a neurotransmitter that we perceive as pleasurable. Sugar molecules hitting your taste buds, for example, send action potential signals to your brain that cause neurons to release dopamine, which we perceive as pleasure. We are social animals, and eye contact and friendly interaction stimulate neurons in the brain to produce oxytocin, which we perceive as a warm emotional bond. Sex, when it's consensual, causes neurons in the brain to release dopamine, which is why sex is fun. Our brains are constantly tracking all the stimuli associated with these good feelings—a dessert menu, the faces of our friends, a flirtatious smile—and we build complex behaviors to seek them out. Advertisers take advantage of this, putting scantily clad models or Matthew McConaughey in their car commercials so you associate Buicks with sex.

We can learn to respond to negative stimuli as well. When we're surprised by a loud bang, or scared by a snake in the garden, or giving a public speech, those stimuli activate our "sympathetic" nervous system, the fight-or-flight response that acts as our body's all-purpose fire alarm. Action potential signals are sent to the adrenal glands to squirt some adrenaline (also called epinephrine) into the bloodstream, to the eyes to widen the pupils, to the gut to pause digestion, to the skin on your palms to produce sweat, and more, all in a coordinated effort to get you out of danger. Like birthday presents from your elderly parents, these evolved responses are well-intentioned and would have made sense a long time ago, even if they are counterproductive now. Damp hands and a faster heart rate would have given your monkey-like ancestors better grip and more oxygenated muscles as they dashed up a tree to safety, but are worse than useless today as you smudge your notes and tremble at the podium for your big speech. As with rewarding behaviors, our brains track the stimuli associated with bad outcomes. We learn to fear and avoid the vodka shots that made us sick or the jerk who bullied us.

All of this highlights that learning is a physical process, a change in the circuitry of the brain. New circuits form as different stimuli and experiences are repeatedly paired. Circuits that are used repeatedly, or associated with an extreme event, are strengthened. Unused circuits get broken apart, their synapses "pruned" like unsightly shrubbery. In childhood, when our brains are busiest with the hard work of learning, they burn more calories each day than any other organ. Each night children go to sleep with a different brain than the one they had when they woke up—the neurons are the same, but the network of connections will have changed. That same process continues throughout our lives, though at a slower rate as we age. Every time you learn something new (say, how brains work), you modify your brain.

Go one step further, and we can understand how behaviors can evolve. Some connections are easier to build than others, and the contours and constraints on learning and behavior are shaped by the prewired circuitry we're born with. Those prewired circuits are built by genes, and as we've seen, genes can change over evolutionary time. Our inborn preferences are there because they promoted the survival and reproduction of individuals that have them. Paleolithic humans who liked sugar were more likely to seek out high-calorie foods that helped them thrive, and if they liked sex they were more likely to reproduce. If conditions change, different prewired circuits might be favored. Cats, for example, have lost the ability to even taste sugar, and have no dopamine response to sweets. Mice are born with a prewired fear response to the smell of cat urine. Species that breed seasonally only seek out sex during a narrow window in the year, when the right set of environmental cues activate those circuits.

One final step, and we can understand how abilities evolve.

The brain isn't just one thing, some general-purpose computer—it's a collection of parts, each dedicated to a different task. There are regions dedicated to visual processing, to interpreting and producing language, to movement, to emotional processing, to building and storing memories, and so on. Most of us end up with the same arrangement: vision processing in the rear, language processing on the left side, and so on. But there's a surprising degree of plasticity, particularly when our brains are young and developing. If a child loses the ability to see, sound- and language-processing networks can take over the unused areas usually occupied with vision. This sort of rewiring probably explains the heightened sensory abilities we seem to find in people like Stevie Wonder or Doc Watson, both virtuoso musicians who lost their vision at an early age.

Similar to that growth in musical ability as sound processing expands into a larger region, other abilities can grow or shrink over evolutionary time as natural selection favors different capabilities. The sizes of different brain regions are determined by the genes that build the brain, and those genes evolve. Cave-dwelling fish that live in complete darkness lose their sight- and visual-processing regions. Bats evolved expanded brain regions for tongue control to produce the clicks used in echolocation. Humans and other primates have large regions of their brains dedicated to visual processing, while dogs have large regions dedicated to interpreting smells.

Humans are smarter than every other species on the planet because of our enormous prefrontal cortex, the brain region used for abstract thinking and problem-solving. All apes have big brains with a sizable prefrontal cortex, but ours are three times larger than chimps, bonobos, gorillas, or orangutans. Only some gigantic animals, like whales, have a bigger prefrontal cortex, but in proportion to body size ours is still much larger.

As I have often been reminded on my forays with the Hadza, being a hunter-gatherer takes an incredible degree of sophistication and intellect. Alleles that build bigger and bigger brains, and in particular a larger prefrontal cortex, have been steadily favored for more than 2 million years, since the dawn of hunting and gathering. As a result, inside our huge, bulbous heads we have more power to think, solve, and create than any other species.

But here is where our string lights analogy fails us, and where modern science runs aground. We've made enormous strides over the past century figuring out how neurons work, how circuits of neurons can control vital functions like breathing and produce complex behaviors, and how our abilities evolve. But none of it has gotten us much closer to what psychologists and philosophers call the "hard problem" of how neurons produce consciousness. How does a tangle of Christmas lights blinking away inside a blob of Jell-O *think*? How can it imagine the fjords of Norway, compose a symphony, or puzzle out a math problem? Your rich inner life full of ideas and dreams emanates from a three-pound mass of buttery soft flesh in your head, and we don't really know how.

That's a remarkable admission. Science has figured out how every organ in your body performs its primary tasks—every organ, that is, except the most important one.

Perhaps someday we'll have an answer to the hard problem. For now, if we want to talk about intelligence, personality, and all the other fun and contentious aspects of how our brains work, we need to step outside our heads and observe. Just as Gregor Mendel figured out the basics of genetic inheritance back in the 1800s by watching plants reproduce, without knowing about DNA or chromosomes, we can watch people in action and try to figure out how their brains are working behind the scenes. Knowing how neurons and reward systems operate will help ground our ideas about how complex functions might work. With the tools of modern genetics, we can even begin to tease apart which aspects are prewired and which are learned.

ARE WE SMART ENOUGH TO UNDERSTAND INTELLIGENCE?

The science of intelligence is a young one. It wasn't until the late 1800s and early 1900s, as the formal educational systems of Europe and the U.S. were developing, that the modern, Western conception of intelligence emerged. Alfred Binet, a psychologist working in France, was tasked by the French government with developing a test to identify children who were falling behind. Working with his student Theophile Simon, Binet developed a set of questions that started very easy and grew progressively more challenging. Children were scored relative to others their age. A child who correctly answered questions that most eight-year-olds get right, but got harder questions wrong, would get a "mental age" of eight. The Binet-Simon test became the first intelligence test in widespread use and remained the basis of intelligence testing for decades.

To compare results from the Binet-Simon test across children of different ages, psychologists hit upon the idea of dividing a child's mental age by their chronological age to calculate an "intelligence quotient," or IQ. One could then multiply that ratio by 100 to give a nice, whole number. If the child was 10 years old with a mental age of 8, they'd have an IQ of 80 (calculated as $8/10 \times 100$). If they were 6 years old with a mental age of 7 they'd have an IQ of 117. Most children, by definition, had a mental age that matched their chronological age, and had an IQ of 100. We still use this convention today, even with updated tests: average IQ is set to 100.

Working with hundreds of children over the course of his

career, Binet viewed intelligence as multifaceted and dynamic. He was hesitant to see a child's cognitive ability boiled down to a single number, and he pointedly rejected the idea that intelligence was innate or fixed. In his view, environment clearly played a big part in a child's IQ score, and intelligence could be improved with environmental enrichment. He saw his test as a tool to identify children who needed help.

Across the Atlantic, U.S. researchers were using Binet's test for a very different purpose. Eugenics, the idea that societies should control marriage and reproduction to encourage "good" genes in the population, was all the rage among the educated elite, including scientists, politicians, and their wealthy benefactors. Evolution was still a new idea, and people were enamored by the thought of helping natural selection along by selectively breeding people like cattle. Some vocal proponents espoused noble goals, thinking eugenics could rid the world of disease and disability. Many were also racist, some of them openly, and they liked the idea of promoting bigger families for wealthy Whites like themselves, while limiting reproduction for Black people and other minorities.

Eugenics was always presented as a serious, evidence-based, scientific undertaking instead of a racist human-breeding program, which was key to its acceptance by the broader public. Its central tenet was that genetics determine nearly everything about you. Environmental influences were ignored or brushed aside. That perspective, wrong as it was, meant they could claim to assess the quality of your genes simply by measuring your traits. All they needed were the right tools to evaluate who had "good" genes. It probably won't surprise you to learn that the people determined to have good genes were invariably White and upper-class, like the people doing the testing.

Lewis Terman, a eugenics zealot and psychology professor at Stanford University, saw Binet's test as the perfect tool to measure intelligence and assess people's genes. He tweaked the exam a bit, creating the Stanford-Binet test, and sold it widely. Soon, IQ tests were in use throughout the U.S. to identify children and adults who were deficient and, in Terman's view, beyond hope. When word got back to France, Binet was horrified to learn what researchers in the U.S. were doing with his test. He died in 1911, before he could do anything about it.

By the 1920s, Terman was a leading voice in the American eugenics movement and regarded as an expert on matters of intelligence. He dismissed the intellectual abilities of entire swaths of the population, pronouncing Italian, Spanish Indian, Mexican, Portuguese, and Black people as intellectually inferior, incapable of abstract thought. His views came to dominate education and other social policy in the U.S., with real-world consequences. If IQ was genetic and unchangeable, as Terman believed, there was no point in trying to educate or help anyone who scored poorly on his test. Federal policies were changed to restrict the immigration of Jews, Eastern Europeans, and others deemed genetically inferior. States passed laws to sterilize people with low IQ scores and lock them away in institutions, to prevent them from spreading their faulty genes. Tens of thousands of people were forcibly sterilized in the U.S. in the early and mid-1900s, more than twenty thousand in California alone.

There were strong voices of protest all along the way, researchers and policymakers who objected both on scientific and moral grounds. Many people knew that environments mattered, just as they knew that the social policies pushed by the eugenics movement were horrific. Their arguments gained little traction. Eugenicists like Terman were always careful to acknowledge *some* role of the environment, before quickly arguing that genes were much more important. Politicians were careful to sympathize publicly with the conditions of the poor and minority communities they were hurting, and to focus on what they insisted was the greater good. Some, no doubt, truly believed in the promise of improving our species. And there were a handful of prominent Black scholars, W. E. B. Du Bois among them, who supported eugenics and helped to provide political cover for clearly racist policies.

But cracks eventually began to show in the scientific foundations of the eugenics agenda. The revolution began in an unlikely place, at an orphanage in Davenport, Iowa. Orphanages throughout the U.S. were busy in the 1920s, thanks in part to eugenic policies that pulled babies away from mothers deemed incompetent. By the 1930s, the Great Depression had made things worse, as families were forced to give up children they couldn't support. Those children often grew up in overcrowded facilities with little enrichment besides bland food and a bed.

Intelligence testing typically revealed exactly what eugenicists expected: many of the children growing up in these orphanages had low IQ scores. The standard explanation was poor genetics. Their low-IQ parents had given them bad genes.

Psychologists Marie Skodak and Harold Skeels weren't so sure the standard story was right. As detailed in Marilyn Brookwood's fantastic book, *The Orphans of Davenport*, both had taken jobs with the orphanage in Davenport, tasked with looking after the children there and finding families to adopt them. In annual checkups after these adoptions, Skodak and Skeels noticed that nearly all the children lucky enough to be adopted into loving families scored in the normal or even "superior" range for intelligence, in sharp contrast to their unlucky peers who had stayed in the orphanage. More troubling, children who entered the Davenport orphanage with normal IQ scores often saw their test scores decline with each month they spent there. The apparent malleability of intelligence alarmed Skodak and Skeels, who had both been trained in the standard dogma of the time, that intelligence was determined by genes. If eugenicists like Terman were right, it shouldn't matter where these children grew up or how long they lived in the orphanage. Their IQ scores should have stayed the same.

The breakthrough came when Skodak and Skeels placed a group of infants from Davenport in the care of an institution for women with exceptionally low IQs. Initially done to ease overcrowding at Davenport, the first two children placed at the women's facility had flourished. Both had behavioral delays and IQ scores below 50 when initially tested as infants at Davenport, indicating severe mental deficiency that would require lifelong institutional care. After just eight months at the women's facility their scores had nearly doubled, and they behaved like typical toddlers in every way. Intrigued, Skodak and Skeels randomly chose another thirteen infants with low IQ scores to place at the women's facility, and another twelve orphans with similar scores who would remain at Davenport. Two years later, every one of the children placed in the women's facility had seen their IQ scores increase, by an average of 28 points. Their unlucky peers who remained at Davenport saw their scores *decline* by the same amount. These differences remained stable through childhood.

The changes in IQ between these groups couldn't easily be explained by genetics. Mothers from both groups had the same low IQ scores. Nor could the jump seen in children at the women's facility be explained as a consequence of high-achieving, high-IQ caregivers. The children were mainly looked after by women housed at the institution, all of whom were there because they had been identified as having very limited intelligence, with IQs typically below 70.

The main difference for these children was a supportive and caring environment, where each had individualized attention

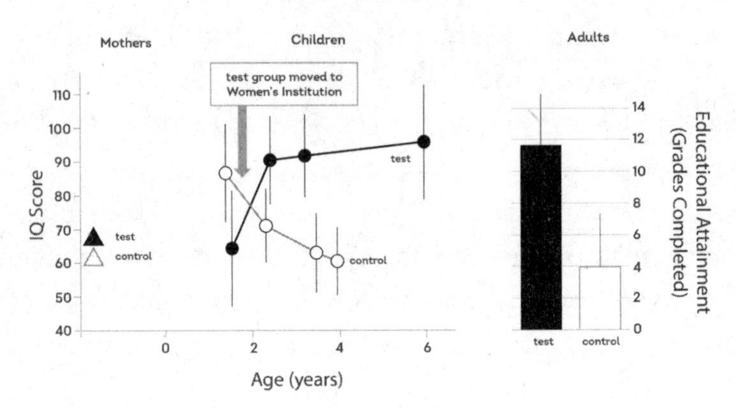

Figure 3.2 Davenport orphans were randomly assigned to two groups; their mothers had similar low IQ scores and both groups of children had low IQ scores at the start of the study. Test group children were raised in an institution for low-IQ women. They quickly increased IQ scores and maintained them throughout childhood. Control group children maintained low IQ scores. The test group children went on to complete an average of twelve years of schooling, while the control group completed only four.

from adults who loved them. The effect on their lives was profound. Children from the women's facility group completed twelve years of schooling, on average, while few of the orphanage children finished elementary school. Thirty years on, when Skeels followed up with the children from both groups in their adult years, all but one from the women's facility group had a steady marriage and were able to hold good jobs like teachers, nurses, or salesmen. Of the group who stayed in the orphanage at Davenport, only one was married. None had what most of us would consider a great career, although some worked menial jobs such as dishwashers or gardener's assistants. Four had remained in institutions their entire lives.

Skodak and Skeels' work became a landmark study in a growing body of research showing intelligence was malleable and heavily influenced by environmental factors, particularly in early childhood. As the consensus shifted in the 1950s and '60s, eugenics fell out of favor. Its foundational assumption, that intelligence and other traits were largely genetic, with little environmental influence, became untenable.

Public awareness of the atrocities of the Nazi regime pushed eugenic ideas further into the sewer. The Nazis were eager students and close allies of American eugenics societies in the 1920s and '30s, adopting their principles and approaches. The connection to Nazi concentration camps exposed the U.S. eugenics movement for the horror it was.

Sadly, much like a stubborn venereal disease, racist biology has continued to hang on in the damp and fetid nether regions of pseudoscience. Today it is typically discussed in terms of "race realism" or other scientific-sounding garbage. The argument takes on different forms, but at its core it is the same as Terman's: IQ scores and other education outcomes differ by race, and that means the low-scoring races are genetically deficient.

Racists even propose clever-sounding hypotheses to explain why groups might differ in intelligence. "Cold winters" hypotheses argue that life in the tropics is so easy that you don't need to be smart, and that northern climates selected for intelligence to survive harsh winters. These people have clearly never tried to track a wounded impala through the savanna. Nor have they navigated to new islands by canoe across the vast expanse of the Pacific like Indigenous Polynesians, learned to thrive in a blank and hostile desert like many aboriginal Australian groups, or mastered any of the other sophisticated strategies common among tropical populations. Nothing screams armchair racist quite as clearly as a guy living a pampered life full of modern conveniences that he barely understands and could never re-create arguing that people smart enough to live off the land are somehow intellectually inferior.

"Life history" hypotheses suggest that some groups are adapted to put more energy into reproduction and less into brains. This tired trope dates back at least to the influential British economist Thomas Malthus in the 1800s and the fear among well-heeled English of Irish overpopulation. If worrying about the Irish doesn't sound racist, it might surprise you to learn that Irish immigrants were not considered "White" in the U.S. until the early 1900s. (Races are cultural categories, not biological ones, and they are as strange and fanciful as human imagination.) Hand-wringing about the fertility of Black, Latino, and other minority communities has been a favorite culture war pastime in White U.S. communities for over a century. Never mind that big families were the norm throughout northern and western Europe and until very recently. George Washington was one of ten children, as was Thomas Jefferson. My German Catholic grandmother had seven children, not unusual for the early 1900s, which is more than the typical Hadza hunter-gatherer mom. Amish and Mennonite communities in the U.S., descendants of White Europeans, still have large families, but no one seems to connect that to intelligence.

One point racists like to harp on is that the racial gap in IQ scores and educational attainment in the U.S. has persisted despite decades of social and education programs to address it. That argument fails to recognize (or willfully ignores) two countervailing facts.

First, early-life educational experience and opportunity still differ among racial groups in the U.S. These differences are apparent before kids even start school. Entering kindergarten, Black children are typically several months behind their White peers in reading and math. The gap can be erased with quality preschool education, but that opportunity is often out of reach for Black families in poor neighborhoods. And, because school funding in the U.S. is tied to real estate taxes, and housing policies have historically kept Black families out of rich neighborhoods, schools with a high proportion of Black students have substantially less money per child than other schools and are more likely to be staffed by inexperienced teachers. Similar disparities affect children in Native American, Latino, and other minority communities.

Second, the IQ scores of *every* group on the planet have been rising for the past century. This surprising observation is called the Flynn effect after the researcher James Flynn, who first recognized it. Because Binet's test has been the basis of IQ testing since the early 1900s, it's possible to compare scores over time. Today, the typical U.S. child taking the test (scoring 100 when compared to his peers) answers more questions correctly than the average kid did a century ago, meaning today's average-IQ child would be scored near "genius" level compared to children the same age in the 1920s. The Flynn effect clearly shows that environments affect IQ—the change is far faster than genetic evolution could have produced. The exact causes are still debated, but likely include a mix of better early education, more stimulating childhood experiences, and improved nutrition. (As we touched on in the past two chapters, nutrition in early life affects brain development.)

If scores for *every* group are increasing year over year, it makes sense that gaps between groups will persist. It's like a marathon in which one group gets a head start: midway through the race, everyone has moved far ahead, but the gap is still there. Interestingly, it appears the growth in IQ scores among wealthy countries may have slowed in recent years, perhaps even declining slightly in some. This could allow others to finally catch up, but time will tell.

Despite the clear problems with using IQ testing to argue for innate, genetic differences in intelligence between groups, the practice persists. Some have even developed "national IQ" datasets to compare intelligence across countries. There are fatal problems with these datasets in terms of sample size and the extent to which they are representative of a particular population, problems that disqualify them from being taken seriously as real evidence.

There's an even larger problem in the way these figures are interpreted. Unsurprisingly, countries with a lower "national IQ" tend to be poorer and have lower standards of living. Like Terman, national IQ proponents argue that these populations are poor because they have a low IQ. But just like Terman, they have the causal arrow pointing the wrong direction. Kids in poor countries don't have the opportunities or advantages of children in rich countries, and IQ scores reflect that.

There is also the inescapable issue of cultural fluency. The Hadza aren't included in these national comparisons, but you can imagine how they'd fare. Without formal, Western-style schooling from an early age, we'd expect low scores on our IQ tests, even though their intelligence is obvious to anyone who spends time with them. And *we* would fare poorly on any IQ test that *they* would develop, as my experiences tracking impala and foraging for tubers make plain. The same limitations affect any comparisons across countries or cultures. Group differences in IQ scores tell us about environments, nothing more.

EVERYTHING NEW IS OLD AGAIN

The revolution in genetics over the past decade, in particular the development of GWAS methods, has reinvigorated research in the genetics of intelligence. Studies with large enough sample sizes generally find what twin studies have shown for decades, that measures of intelligence like IQ scores and educational attainment (the years of schooling a person obtains) are heritable. The effect appears to be modest, with GWAS studies generating heritability values in the neighborhood of 0.10 to 0.20. Twin and family studies give higher estimates, but those studies have a harder time separating out environmental effects and are prone to overestimating the role of genetics. Still, it seems clear that the alleles you get from your biological mom and dad influence the way your brain is built and wired, which affects to some small degree how well you score on IQ tests and the level of education you pursue.

Other mental and behavioral traits are heritable as well. Many psychiatric disorders, including schizophrenia, bipolar disorder, and depression, have heritabilities similar to those for IQ. Studies of the so-called Big Five personality traits—neuroticism, extroversion, openness, agreeableness, and conscientiousness—suggest the genetic influence is measurable but modest, with heritability from GWAS studies of around 0.10 to 0.20 (again, twin studies give higher estimates). Something about the alleles we carry sets up our reward and behavioral circuitry in subtle ways that tend to make us more cautious or curious, outgoing or reserved, messy or neat. Some studies have even pointed to likely mechanisms, drawing connections between genetic variants associated with personality traits and variation in synapse and neurotransmitter function.

We also find genetic connections among mental traits. Genetic variants associated with higher IQ and educational attainment are negatively associated with alcohol addiction and attention deficit disorder, for example. Alleles associated with agreeableness and neuroticism are also linked with gambling addiction. Many of the alleles associated with major depression are also connected with schizophrenia, bipolar disorder, and anorexia. Again, these connections make sense given what we know about how the brain works. Neuron circuits and neurotransmitters are shared across tasks and behaviors, and issues in one domain might mean you're more likely to have issues in others as well.

There's even evidence that different kinds of intelligence, like musical ability or athletic talent, are influenced by the alleles we carry. Musical abilities like perfect pitch (the ability to tell that a certain sound is an F-sharp or an A, for example) and rhythm are heritable, as are endurance capabilities and strength. Those crude measures don't quite get at the question of *genius*—surely the genius of Beethoven or Abby Wambach is more than just the ability to identify tones or run up and down the field—but it sure seems like we're moving toward an understanding of the genetics of intelligence across multiple domains, why some kids love construction while others are drawn to calculus. Binet would be proud.

These advances make some people nervous, and it's common to see the publication of a GWAS study on IQ scores or educational attainment ignite intense discussion about the meaning of the results. For racists, they must feel like a vindication, growing evidence that genetics reigns supreme and that, eventually, the consensus will return to Terman's view that racial differences in intelligence are real and genetic. Others point to the many limitations of these studies and reject the idea that they tell us anything new or meaningful. Any parent with more than one child knows that we come into the world prewired with a unique blend of personality and talent. *Of course* genetics plays a role in shaping our abilities. That doesn't mean environment and luck aren't even more important, or that differences among groups are due to genetics.

The back-and-forth on social media and elsewhere can feel like we're back in the mid-1900s, with the nature-nurture debate over intelligence at full throttle. I see it differently. Just as the stern of a sinking boat rises up out of the water before submerging (a hard lesson that Leonardo DiCaprio and Kate Winslet taught us in *Titanic*), I think these high-profile discoveries mark the beginning of the end for gene-centric views of intelligence. Rather than revealing the strengths of the eugenics perspective, they've exposed its weakness. Here's what we know.

First, the heritability estimates for intelligence and behavioral traits calculated from GWAS and other DNA-based studies are embarrassingly low. Have a look at figure 3.3. When heritability for a trait is *really* high, north of 0.80 or so, it's reasonable to claim that someone's alleles can predict that trait. But those cases are exceedingly rare—so rare that there are few real-world examples. Height, one of the best-studied examples of trait heritability, only has GWAS heritability values of ~0.50. At that level, genetics might provide an educated guess about someone's height, but you wouldn't want to rely on their genetics to buy them clothes.

At a heritability of 0.15, a typical GWAS estimate for IQ, genetics are useless for predicting an individual's score. For any set of alleles, or "polygenic risk score" for intelligence, there will be lots of people with high IQs, low IQs, and everything in between. Sure, for a thousand people with a particular set of alleles you could predict their average IQ score, but that would be little more

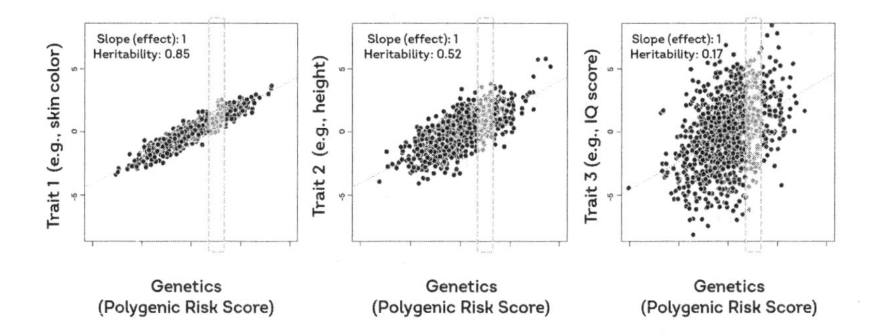

Figure 3.3. Three illustrations of the relationship between genetics (e.g., polygenic risk score) and traits (e.g., skin color, height, or IQ) at different heritability scores. All three have the same slope of 1, meaning the average value for the trait at a given genetic "score" is the same for all three traits. However, the predictive power of genetics for these three traits differs greatly. When heritability is ~0.80 or greater (left panel), genetics predicts the trait quite well. Nearly everyone with a polygenic score of 1 (indicated with the gray dashed box) has a trait value near 1. An example of this in humans is skin color, which is well predicted by genetics. At a heritability of ~0.50 (middle panel) the effect of genetics on a trait is visible, but there is clear variation in the trait that is not explained by genetics. People with a polygenic score of 1 have trait values that range from −2 to 4. A classic example of this degree of heritability is your height. When heritability is ~0.15 (right panel), a typical value for IQ in GWAS studies, genetics has no predictive value for individuals. People with the same genetics will differ wildly in the trait, and people with the same trait will differ wildly in genetics. Among people with a polygenic score of 1, some have among the highest trait values, while others have among the lowest.

than a nerdy parlor trick. The variability in IQ among people would be so high that you wouldn't be able to make useful predictions for individuals or small groups. When heritability is this low, environmental influences swamp genetic effects.

Second, even if we accept the higher heritability estimates from twin and family studies, the genetic association of IQ and educational attainment with other, seemingly unrelated traits undermines the notion that these studies capture anything fundamental about cognitive ability. For example, alleles associated with IQ and educational attainment are also positively correlated with the ages that people first have sex and start having children, and negatively associated with attention deficit hyperactivity disorder (ADHD). It's possible that people with exceptionally smart brains *also* tend to delay sexual activity and are protected from ADHD. But it seems far more likely that kids who are less distracted by sex and able to sit through hours of class without complaint are more likely to do well on IQ tests and seek out more schooling. IQ tests and educational attainment probably measure a range of traits that have nothing to do with the ability to remember or process information, develop creative ideas, or any of the other things we associate with being smart.

IQ and educational attainment also measure the ways society treats us. As we saw in the last chapter, tall people tend to have higher IQ scores and finish more years of schooling, even though being taller does not improve brain function. The relationship must be due instead to the unconscious and subtle ways people treat taller children, or an environmental signal that some kids are getting the nutrition they need to grow tall and nourish their growing brains. In the U.S., alleles that affect skin color would have strongly predicted the ability to attend school just a few decades ago, and they still predict educational opportunities today. While GWAS and twin studies try to minimize these issues, they are still vulnerable to including hidden environmental effects in their calculation of heritability, inflating their estimates of genetic influence.

And finally, remember that heritability values only apply to the population in which they're measured, and have nothing to tell us about differences between groups. In fact, the idea that human groups, whether "races" or other denominations, might differ in the genetics of intelligence completely misunderstands our species' evolutionary history and the way brains work. As we discussed in the last chapter, most of the biological diversity we see in our species is random and neutral, while a small proportion of variation is due to local adaptation. We can place the few alleles that *aren't* shared across groups into those two broad categories: neutral or local. Neutral alleles, like those that affect most aspects of our face shape, have no effect on survival and reproduction. Population differences in neutral alleles can persist because natural selection ignores them. Local alleles, on the other hand, are genetic differences between groups that reflect local adaptations to the particular challenges of specific, local environments, like we saw with body proportions and climate. Population differences in local alleles persist because different alleles are advantageous in different environments.

For differences in intelligence among groups to be due to genetic differences, the alleles involved would need to be either neutral or local—neither of which is possible.

Alleles that affect intelligence could never be neutral because intelligence is too important. Our lineage has been under strong selection for bigger, smarter brains for more than 2 million years. As sophisticated cultural strategies came to dominate every aspect of our lives, intelligence became essential for both survival and reproduction. Any allele that improved the ability to think, create, deceive, or learn would have spread like wildfire.

And alleles for intelligence couldn't be local because the ecological pressures for smarts don't differ among populations. Adaptive differences between groups can only evolve when ecological pressures are strong *and* consistently different between regions. Skin color, as we'll see in chapter 8, is an adaptation to variation in ultraviolet light intensity, which is consistently high at the equator and low at the poles. Body proportions are an adaptation to consistent temperature differences between equatorial and arctic settings.

Selection pressures for intelligence are strong, but they're the same for everyone around the globe. They aren't local. No matter where you're born, your brain has the same job, to absorb the cultural and ecological knowledge of your community. That process takes nearly twenty years, whether you're born in the Hamptons or Hadzaland. And it's a moving target, changing rapidly as technology and culture develop and habitats and climates evolve. The norms you learned and the problems you faced might be completely different from those of your parents (or your kids). Staying nimble requires our brains to be born unfinished and plastic, ready to learn, without too many of the specifics prewired. Every culture is complex, every setting challenging. Life is never simple or static.

Natural selection hasn't been—*can't* have been—working on the *content* of our brains: our vocabulary, the facts we internalize, the problem-solving strategies we use, the patterns we learn to recognize, or the linkages between cause and effect we come to anticipate. All of that content changes too quickly for biological evolution to hardwire into our brains. Instead, like software, it must be learned and uploaded as we grow. Natural selection can only shape the hardware—the capacity to learn, the ability to anticipate—and that hardware has been under the same selection pressures in every population around the globe for eons. Under those conditions, we would not expect intelligence differences between groups to evolve, and we have no evidence that they ever have.

Despite claims to the contrary, IQ tests are content tests. They test the cultural software you've uploaded and absorbed. The only way that IQ tests can ever be useful for assessing people's mental hardware is in very limited cases where everyone in a group has been exposed to the same content, under the same conditions. IQ tests for a classroom of children who are all the same age and have had very similar upbringings *might* be able to tell you something about the way they're wired, but as we've seen with the Davenport orphans, even those differences are labile. Once we start comparing across groups, or between individuals with different backgrounds, environmental differences in content and the manner in which it's provided quickly obliterate the ability to draw any conclusions about genetic differences or innate mental abilities.

IMPALA QUOTIENT

We found the impala eventually (if you were wondering). Hadza intelligence prevailed, piecing together a set of subtle and confusing clues to win the day and feed their families. I still don't know how they did it.

I'm not a very high-IQ Hadza man, and I'm okay with that. I do all right in my own community, which is what my brain is built to do. My intelligence, like yours, came together piece by piece throughout my development, a rich cultural inheritance woven atop and through a bare-bones framework of neural circuitry that was shaped by the particular set of alleles I was dealt. Separating the environmental influences from the genetics would be impossible now, with each embedded in the other. Genes helped to sketch out the initial scaffolding, tilted my temperament in a particular direction, but the content of my mind, including the stuff

I know and the ways I express myself, are products of my environment. My neural circuitry and cognitive abilities have been shaped and reshaped over nearly five decades of learning through the interplay of my unique set of alleles and experiences. Synapses connected or detached, networks strengthened or atrophied as I absorbed the lessons from my community. If I had grown up in Hadzaland, or a Davenport orphanage, or a UN refugee camp, I would very literally have a different brain. It's the same for you.

Despite decades of research and the loud insistence of racists, no genes associated with intelligence have ever been found to differ between groups of humans. I don't think they ever will—the pressures on intelligence are so strong and so similar around the globe that we shouldn't expect groups to differ in intelligence any more than we'd expect them to vary in the number of fingers and toes. But if researchers ever do find subtle genetic differences in intelligence between groups, I hope the racists are ready. The chances that melanin-challenged pseudo-intellectuals coddled by modern technology hold any unique alleles for intelligence seem remote. If there are any groups that do, my money is on the populations that have continued to thrive through sheer grit and guile, without the padded corners and safety nets of the industrialized world. I just hope the Hadza are cool about it.

THE ENVIRONMENT SHAPES ALL OF OUR ORGANS, BUT ONLY BRAINS have returned the favor. Our ancestors have been altering our landscapes for hundreds of thousands of years. With the development of farming, towns, and eventually cities, we completely reimagined what our worlds would look like. Industrialization accelerated this radical transformation, creating electrified ecosystems of concrete, glass, and steel where many of us spend our lives. Every facet of daily life, all the knowledge embedded in our phones, architecture, medicine, computers, and cars, is the result of a long, glacial process of cultural accumulation passed down in an unbroken chain from the Paleolithic through today. Campfires became nuclear fusion. Stone tools became spaceships.

As lovely as our modern world can be, there's a monkeys-running-the-zoo element to it that should give us pause. We've engineered the world to our liking, putting delicious food at our fingertips, building cars, elevators, and other machines to move our passive bodies from place to place and perform our daily chores. Most of this development has been unintentional, driven by market forces, technological developments, and population growth. To paraphrase the mathematician and chaos theorist Dr. Ian Malcolm, we were so preoccupied with whether or not we *could*, we didn't stop to think if we *should*. But by changing our environments we've also changed ourselves, in ways we're only beginning to appreciate.

4

HEART AND AIR SUPPLY

Prediction is a tricky business in biology. As we see with all the excited headlines about the genetics of intelligence, when you get right down to it it's awfully difficult to know how life will turn out for any of us. The interplay of genes, environment, and chance is a complex dance, and we live in an unpredictable world full of unpredictable people. Guessing what's coming, even a few steps ahead, is a challenge.

But I can make two solid predictions about your body, even without knowing anything about you—predictions I'd be willing to put money on. The first is that it won't last forever. At some point, the protein robot you've been driving around like a damp RV will fail, and you'll die. Mine will, too. Of this I'm certain. The long-term mortality rate for humans has been holding steady at 100 percent for as long as anyone has been paying attention.

My second prediction is that your cause of death will be a culprit you probably never saw coming: evolutionary mismatch. Your body will break down and fail because it is evolved to thrive in an environment that's radically different from the one in which you live. That mismatch will lead to obesity, heart disease, diabetes, cancer, or some other problem that will ultimately cause your demise. Sure, it's possible you'll die from an infection, accident, or some other non-evolutionary cause, but the smart money is on mismatch diseases, which have grown to become the leading killers worldwide. If you're reading this book, chances are good that you live in an industrialized society, which increases your odds of a mismatch disease even further.

It wasn't always this way. A few generations ago, before most of the world transitioned from farms to factories, heart disease, diabetes, and other mismatch diseases were minor players in public health. Then, in just a couple of centuries, industrialization and modernization radically transformed the ways we live and the ways we die. There's been a lot of good, to be sure. Public sanitation, soap, antibiotics, vaccines, and modern medicine have made life longer and better. Today, the average American lives two decades longer than they did just a century ago. But there have been costs as well. Life in the developed world saddles us with a range of problems we never had before.

No part of our body has been hit harder than our cardiovascular system. Along with the lungs, the heart and blood vessels serve the vital function of delivering oxygen and other nutrients throughout the body, but that's not all. They also help to clear waste, keep our blood acidity in check, and give us the power of speech. They've adapted to pull oxygen from thin mountain air and teamed up with the spleen to help people forage underwater. And these organs never sleep, racking up over 2 billion heartbeats and half a billion breaths by the time we reach our seventies. Yet it's these same rugged and adaptable organs that are most likely to let us down as we age.

We take it for granted that our hearts and lungs will wear out in our golden years, our blood pressures creeping upward and arteries hardening as our hair turns gray, but those changes are far from inevitable. Communities like the Hadza, with lifestyles similar to our hunter-gatherer ancestors, seem to be immune from heart disease even into their sixties, seventies, and beyond. How do our hearts and lungs work, and what can we learn from populations like the Hadza about keeping them healthy?

PUMP IT

Your heart is a fist-sized pump made out of muscle and nerves. There's a right and left side, each with an "atrium" above and "ventricle" below, chambers that expand and contract to pull blood in and squirt it out. Blood enters the right side of the heart, pouring into the right atrium from two massive vessels the diameter of your thumb, the superior vena cava coming down from the neck and shoulders, the inferior vena cava coming up from the liver. After it fills, the atrium contracts just as the ventricle below it relaxes and expands, and blood pours into the open right ventricle through a trapdoor in the floor of the atrium. Now the ventricle contracts, but the trapdoor to the atrium (known as the tricuspid valve) is closed, and the blood has only one place to go: into the large vessels of the "pulmonary trunk," which distributes the blood to the lungs.

The arteries that feed the lungs branch into smaller and smaller vessels, eventually becoming so small that the blood cells squeeze through single file. These ultrathin vessels, called capillaries, wrap around little air sacs in the lungs, structures called alveoli that look like miniature bunches of grapes. The walls of capillaries and alveoli are each just one cell thick, one-tenth the thickness of a human hair. Oxygen molecules from the freshly inhaled air in the alveoli happily jump across those walls and onto the red blood cells traveling past. Carbon dioxide jumps the other way, from the blood into the little pocket of air in the alveoli, to be exhaled and discarded.

After passing by the alveoli the capillaries join back up with others, forming larger and larger vessels that will eventually exit the lungs. Oxygen-rich blood from the lungs now enters the left atrium of the heart. Just as we saw on the right side, the left atrium contracts as the left ventricle below it opens. Blood rushes into the left ventricle through the mitral valve in the floor of the left atrium. The left ventricle then contracts, and with the mitral valve now closed the blood has only one exit, into the aorta, the largest and strongest vessel in your body.

The aorta arches up and left,* with thick arteries spraying out from the top of its arc to feed the head and arms. It then plunges downward through the torso, with arteries branching out to service the liver, kidneys, intestines, and other organs. At about the level of your navel it splits to form the left and right femoral arteries that run down through your groin to feed your legs.

As arteries leave the aorta and burrow into every corner of the body, they branch down to ever smaller vessels, just as we saw in the lungs. Capillaries weave through every bit of tissue, leaving no part untouched, no cells unserviced. Oxygen moves easily across the thin capillary walls from the fresh blood into the hardworking cells. Carbon dioxide moves from the cells into the blood, to be dumped off in the lungs. The capillaries join up again, forming larger and larger vessels. All veins eventually feed into the inferior

---

* Occasionally, the aorta rips apart spontaneously, an unfortunate event called "aortic dissection." Like a broken water main, the heart pumps all the person's blood into their chest cavity within minutes, typically with dire consequences for the unwitting victim. *One star. Would not recommend.*

or superior vena cava, dumping their oxygen-poor and carbon-dioxide-rich blood into the right atrium for another lap.

Circulation of the blood was a mystery to the ancients, who generally held that the heart was the seat of consciousness and the soul. That makes sense—it's one of the few places in the body you can hear and feel activity, and your heart rate corresponds with your state of mind. Who would have guessed the inert lump of fat in your head is where the real action is?

The earliest scientific attempt to understand the heart's function is typically credited to Galen of Pergamon, a Greek physician and philosopher who was also the personal doctor of the Roman emperor Marcus Aurelius in the second century CE. From the bodies of dead gladiators and experiments on living pigs and monkeys, Galen figured out the heart pumped blood and came to understand the basics about valves and vessels. But he completely missed the boat on circulation. Galen thought blood was produced by the liver, then pumped by the heart and vessels out into the body and consumed by the tissues. He was also certain that blood from the right side of the heart passed through to the left side through tiny holes in the wall that separates them. His mistaken views on the heart* would remain the consensus for more than a thousand years.

Scientists in both the Near East and Europe eventually figured

---

* In the womb, your blood *does* pass directly from the right side of the heart to the left. Oxygenated blood enters the fetus from the umbilical cord and connects to a large vein below the liver, eventually dumping its blood into the right atrium. The lungs (which aren't breathing anyway) are bypassed, and blood passes through a hole, the "foramen ovale," directly to the left atrium. The foramen ovale closes immediately after birth as the baby takes its first gasping breaths. In some people it never *quite* closes completely, but this rarely causes trouble—the hole is typically very small and very little blood crosses from the right to left atrium. Maybe Galen was misled by dissecting a dead gladiator with an unclosed foramen ovale?

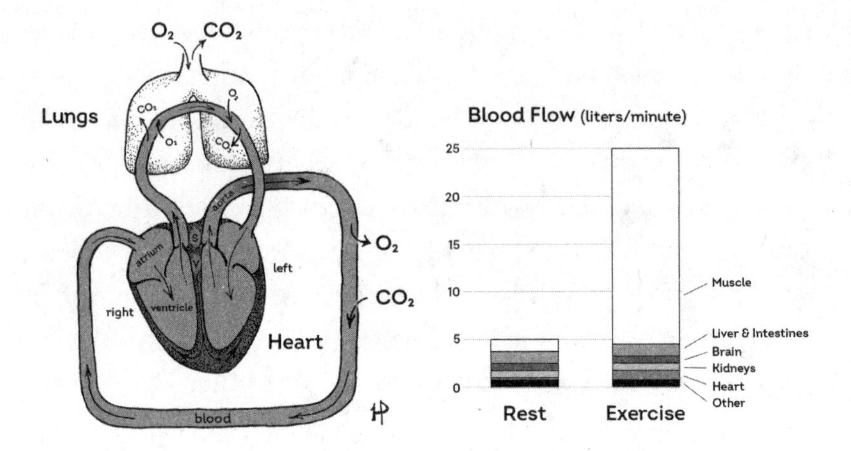

Figure 4.1. Blood flow through the heart and lungs. Blood flows from the right atrium to the right ventricle, then up through the lungs, where it picks up oxygen (O_2) and discards carbon dioxide (CO_2). Then it flows through the left atrium and into the left ventricle, where it's pumped through the aorta and out to the rest of the body. The heart's pace is set by an electrical impulse that starts in the SA node (S) and then passes through the AV node (V), down between the ventricles, then excites and contracts the ventricles. At rest, the heart pumps five liters of blood per minute, and most blood (and oxygen and nutrients) flows to the brain, liver and intestines, and kidneys. During exercise, cardiac output can rise to twenty-five liters per minute or more and most goes to the muscles.

out that Galen was wrong about blood passing from the right to left side of the heart, realizing instead that it circulated through the lungs. But it wasn't until the early 1600s that William Harvey, an English anatomist, pieced the rest of the puzzle together. The blood pumps about five liters of blood per minute at rest, and twenty-five liters (more than six gallons) a minute or more during intense exercise. Using rudimentary measures of this cardiac output, Harvey argued there was simply no way that blood could be produced by the liver and consumed by the tissues so quickly. Instead, he surmised, it must be circulating through the body and returning to the heart. In fact, every drop of blood in your body circulates at least once per minute, and much faster when you're active.

Harvey didn't live long enough to find out where blood is produced—he died in 1657, nearly a decade before the discovery of cells and centuries before we knew where blood cells formed. We know now that blood is a roughly fifty-fifty mix of water and red blood cells, with some white blood cells (immune system cells), proteins, and other bits in the mix as well. The water mostly comes from food and drink. Both red and white blood cells are made in the bone marrow, soft spongy tissue that resides in the hollow centers of your bones, at an astonishing rate. You make 300 *billion* each day, more than 3 million every second.

Red blood cells are chariots for oxygen molecules. They've each got millions of "hemoglobin" molecules inside, proteins with four iron atoms embedded in them. Those iron atoms grab on to oxygen molecules as the red blood cells pass through the lungs. Iron deficiency in the diet is one common cause of "anemia," a condition in which the body can't make enough hemoglobin or red blood cells, causing fatigue and other problems from a lack of oxygen delivery. Iron gives your blood its distinctive red color, and it's brightest in the arteries when oxygens are stuck to all the iron atoms. Blood in your veins is darker, having lost some of its oxygen to your hungry cells.

It wasn't until the late 1800s and early 1900s that anatomists began to understand what made the heart beat. A small cluster of nerve cells called the sinoatrial node, or SA node, sits embedded in the upper wall of the right atrium.* It's connected by neurons

* Fans of human evolution might be interested to learn that the SA node was discovered by Sir Arthur Keith, a British anatomist and major player in the early days of human fossil research. Keith's racism (he was a proud eugenicist) made him easy prey for the Piltdown hoax, an infamous episode in the early 1900s in which *someone* (probably Charles Dawson, who was posthumously discovered to

to a second cluster, the "atrioventricular node" or AV node, which sits between the ventricles at the bottom of the right atrium. Neurons project from the AV node down between the two ventricles to the very tip of the heart before curling back and fanning out across the walls of the ventricles like ivy on an old church.

The neurons in the SA node are similar to ones we discussed last chapter, but with a critical difference: they keep a beat. Instead of waiting to be stimulated like normal neurons, they generate their own action potentials on a regular rhythm . . . *bang*, *bang*, *bang*. Action potentials from the SA node cause the atria (plural of "atrium") to contract, squeezing blood into the ventricles. They also travel to the AV node, causing it to send out action potentials of its own. Those signals zip down between the ventricles and spread out across their thick muscular walls, causing them to contract. Then the cycle starts again . . . SA node *bang!* atria contract, AV node *BOOM!* ventricles contract . . . in a never-ending rhythm that lasts your whole life.

All this electrical activity can be a source of trouble. SA or AV node problems can cause irregular heart rhythms like atrial fibrillation (A-fib) or ventricular fibrillation (V-fib). Heart attacks, caused by clogs in the thin coronary arteries that deliver oxygenated blood to the heart itself, can damage the muscle or neurons and mess up the electrical signaling or muscle contraction. And of course, being electrocuted is bad news because it can derail the normal signaling entirely. Paradoxically, an electrical burst from an automated external defibrillator (the emergency AED kits you see at airports and elsewhere), the paddles you see doctors use in hospital shows ("Clear!" *Bang!*), or an implanted pacemaker can

---

be a serial forger) doctored up some human skull fragments and orangutan teeth and passed them off as fossils of a human ancestral species. Keith bought into the scam, certain that humans must have originated in Europe. He even dedicated a stone memorial to Dawson, and went to his grave rejecting the growing fossil evidence of our African origins. Keith was right about the SA node, though.

discharge the heart's electrical system and reset it, letting the SA node take over again, *bang, bang, bang* . . .

Notice that the heart's electrical system doesn't need signals from the brain to beat. It is self-contained, and just like that classic scene from *Indiana Jones and the Temple of Doom*, a heart ripped fresh out of a person's chest would continue to beat on its own quite happily until it runs out of oxygen (but you still shouldn't do it . . .). There are, however, nerves from the brain to the heart that control the pace of the heartbeat. Action potentials from the sympathetic side of the nervous system (the "fight or flight" response) make the heart beat faster, as does the adrenaline produced when we activate our sympathetic system. Action potentials from the parasympathetic side ("rest and digest") slow it down. Parasympathetic signals to the heart are carried by the vagus nerve, which is why vagus nerve stimulation from stressful events would send my dissection partner (whom we met in the introduction) crashing to the floor. Her heart would slow down so drastically that her brain was deprived of oxygen, and she'd pass out.

Control of the lungs works completely differently. There's no muscle in the lungs themselves to inhale and exhale—they're just elastic sacs that inflate and deflate passively like balloons. Slicing one open is like cutting into a pink, wet sponge (or a gray, wet sponge, if you're a smoker). All the little pockets are alveoli, each connected to a branching system of air tubes called bronchi that begin from the trachea (windpipe) in your throat. The lungs sit atop a flat sheet of muscle called the diaphragm, which moves down when it contracts and back up when it relaxes. The up-and-down movement of the diaphragm acts like a piston. Air is sucked into the lungs when the diaphragm descends and pushed out again when it rises. When you're breathing heavily, the muscles in your abdomen, between your ribs, and even in your neck help as well, expanding and contracting your rib cage and the lungs within.

Breathing requires nerve signals from the brain to the diaphragm, action potentials that arise from neurons in medulla oblongata at the base of the brain. The brain is constantly monitoring carbon dioxide levels in the blood, and if they get too high it signals the diaphragm to breathe deeper and faster. Too much carbon dioxide in the blood makes it harder for cells to off-load their waste, and it increases the blood's acidity, which can be toxic.

Unlike your heart rate, you have a good amount of conscious control over your breathing, but prepare to battle your brain stem. Go ahead and hold your breath, see how long you can go. That fire alarm feeling that begins to creep in, pushing you to take a breath, is your brain stem freaking out. Carbon dioxide levels are rising to levels it finds intolerable. Eventually it will wrestle the controls away from you and force you to breathe. When you train yourself to hold your breath for longer periods, you're actually training your brain stem to ignore the rising carbon dioxide signal.

LEARNING TO EXHALE

The sophistication and control of the heart and lungs can make the system seem like a jewel of evolutionary perfection. But evolution is a tinkerer, a junkyard mechanic solving problems with the materials at hand. Trade-offs and limitations are inevitable. Just ask Jimi Hendrix.

Hendrix was a guitarist of otherworldly talent who revolutionized rock music in the 1960s. He was also an avid participant in the recreational chemistry of the era, indulging heavily in a range of legal and illicit pharmaceuticals. On September 18, 1970, in a hotel in London, after taking roughly eighteen times the recommended dose of sleeping pills after an evening of drinking, Hendrix died. But while the drugs were certainly responsible for his death, it wasn't the chemicals per se that killed him. Instead, having passed out and vomited from the massive overdose, Hendrix fell victim to a much more common killer. He choked.

Humans are uniquely vulnerable to choking.* More than five thousand die that way each year in the U.S. alone. Other species don't have this problem, which is fundamentally a plumbing issue. Your larynx (also called a voice box) is the doorway to your lungs. It's a stiff cartilage cylinder that can be closed off at the top by two fleshy lips called "vocal folds" and a flapping lid called an "epiglottis." The human larynx sits in a precarious position, low in the throat, practically begging to be clogged with every bite of food or gulp of water. Why would evolution favor such a dangerous position for the larynx, threatening our breathing and access to oxygen, when every other animal (including our ape relatives) has theirs sensibly tucked up high and out of the way, behind their nose?

It turns out the dumb position of our larynx is the result of evolutionary tinkering to our breathing system to produce language. The sound of your voice is produced by squeezing air through your larynx with the vocal folds pushed together. This is similar to the way a trumpet player makes a *ptbtptpbptptp!* noise by pushing air through their pursed lips (what I'd call a raspberry and my children insist is a fart sound). The *puff puff puff* of air that escapes becomes pressure waves that travel through the air, which our ears register as sound. Higher or lower notes are achieved by pulling the vocal folds tighter or relaxing them. (Testosterone thickens the vocal folds, which is why men tend to have lower voices.)

You form that sound into vowels by manipulating the shapes of your mouth and throat, and cut it into consonants with your teeth, tongue, and lips. The low position of the larynx makes this

---

* Just ask Buffalo Bills fans.

possible. If it's higher up, at the same level with the nostrils as we see in other apes, you could make noise, but the ability to shape that sound into words would be severely limited. That's why it's nearly impossible to get a dog, chimpanzee, or other mammal to form speech-like words. They can still communicate, of course, with a bark or a grunt, but the rich sonic landscape of human language is out of reach. Our ancestors were so social, so cooperative, that the evolutionary benefits of better communication outweighed the increased risk of choking to death. Choking is the price we pay for the ability to speak.

Other adaptations to our breathing and circulatory systems come at a cost as well. When we travel into the mountains, we're faced with the challenge of extracting enough oxygen from the high-altitude air. The evolved solution is to produce more red blood cells. When the liver and kidneys sense low oxygen concentrations in the blood, they produce the hormone EPO, which stimulates the bone marrow to crank out more red blood cells. (That's why some endurance athletes cheat with EPO injections—it gives them extra red blood cells and oxygen-carrying capacity.) It's a good solution, but it increases the ratio of cells to water in the blood, making it slightly thicker. That, in turn, can cause altitude sickness, which typically involves headaches and nausea, but can progress to dangerous and even fatal fluid buildup in the lungs and brain.

Native populations in the Andes, the highest mountain range in South America, live with elevated red blood cell counts their entire lives. They have larger lungs and rib cages as well, through what appears to be a combination of genetic adaptations for increased air exchange and the environmental pressures of growing up at high altitude. But while a number of genetic adaptations to altitude have been identified in Andean groups, they still struggle with altitude sickness. Approximately 15 percent of adults experience chronic mountain sickness. The physiological solution to low oxygen levels carries a steep price for many.

Intriguingly, altitude sickness isn't as much of an issue for native high-altitude communities in the Himalayan Mountains of Asia. Himalayan and Andean populations are descended from different lowland groups thousands of miles and thousands of years apart. Their movements into the mountains were completely independent, and the adaptations they evolved solved the same set of challenges, but in different ways. Himalayan populations carry a particular allele of a gene called *EPAS1* that's involved in the production of red blood cells. This Himalayan allele has the effect of keeping EPO levels and red blood cell numbers low, allowing people to live with the chronic stresses of altitude without developing mountain sickness. This solution comes with its own downsides, as it also means their ability to carry oxygen is limited, but other adaptations in their vessels and breathing rate maintain oxygen delivery throughout the body.

Even more remarkable than the Himalayan *EPAS1* allele is the story of how they got it. As our ancestors spread out across Africa and then Eurasia over the past two hundred thousand years or so, they encountered other closely related humanlike species, like Neanderthals in the Near East and Europe. And, like humans everywhere throughout history, some of our ancestors weren't particularly picky, and slept with them. Our species were so genetically similar that these couplings produced fertile children, hybrids of our species and others. (Some would argue that we should consider Neanderthals and other groups *human* because of this ability to interbreed—a semantic argument that's fun to have over drinks with an anthropologist.) We can find the genetic evidence of these affairs scattered around our genome today, fragments of DNA from other species that allow retail genetics companies to calculate how much Neanderthal DNA you carry, for example. I'm a bit less than 2 percent Neanderthal, genomically speaking.

Most of these fragments don't have any impact on how our bodies function—they're just mementos from our ancestor's wild affairs, like misspelled tattoos from some Paleolithic spring break, and a reminder that humans will sleep with just about anything. Using the distinction we discussed in the last chapter, these alleles would be considered neutral.

The Himalayan *EPAS1* allele is a clear exception. That allele appears to have entered the human gene pool through a Paleolithic tryst with a group called the Denisovans, somewhere in Asia, roughly fifty thousand years ago. For tens of thousands of years it was just there in the mix, a neutral allele that had no strong effect on survival or reproduction. But around nine thousand years ago, as some of those populations started pushing farther and farther up into the mountains, that allele proved to be advantageous. Those with the Denisovan variant for *EPAS1* were free from altitude sickness, and better able to thrive and raise families in the high mountains. It went from neutral to local and became the predominant allele in Himalayan populations, the adaptive *EPAS1* allele we see in virtually everyone native to the Himalayas today.

Another remarkable case of local cardiovascular adaptation was discovered just recently, in a population known as the Sama (also called the Bajau). The Sama live on houseboats in the ocean around the Philippines, Indonesia, and Malaysia, spending nearly all of their lives at sea. Theirs is a hunter-gatherer lifestyle, but in the ocean: they spearfish and collect food in the depths, sometimes more than two hundred feet below the surface, swimming or using weights to hold themselves down as they walk the seafloor. Like many Indigenous groups, their lifestyle is rapidly changing, but traditionally they could spend four or five hours per day underwater, foraging. It's a lifestyle they appear to have maintained for thousands of years.

Life spent partially underwater poses similar oxygen-delivery challenges as life in the mountains. One evolutionarily ancient response to diving, common among mammals, is to contract the spleen, an organ the shape of a child's slipper tucked up high in the left side of your abdomen, beside your stomach. The spleen is a monitoring station for the immune system, a spongelike organ that checks the blood for bacteria and other nasties. Since it's normally full of blood, it's essentially a reserve tank of red blood cells. When you dive into cold water, the spleen contracts, ejecting its payload of red blood cells to help oxygenate the rest of your body. If you train breath-holding, your spleen will grow to do this job more effectively. High-mountain groups, like those in the Himalayas, have larger spleens than lowlanders, apparently from a combination of genetic adaptation and a life spent at altitude.

Natural selection has favored an allele of the *PDE10A* gene that increases spleen size in the Sama, with nearly double the average volume for those carrying two copies of the allele compared to those with none. Other diving-response genes appear to be under selection in this population as well. Environment still matters—all that breath-holding also helps them increase the size of their spleens. But it's a clear case of genetic adaptation, with natural selection responding to a consistent, strong, and localized challenge in the Sama population.

As remarkable as these high-mountain and seafaring adaptations are, they are limited to a relatively small number of humans today. The biggest and most widespread evolutionary changes to our breathing and circulation systems are those to enhance endurance, which are shared across our entire species. In the words of the great evolutionary anthropologist Bruce Springsteen, all of us are born to run.

Our genus has been evolving as hunter-gatherers for more than 2 million years. It's a physically demanding way to make a living. Hunter-gatherers living today, like the Hadza, rack up more minutes of physical activity each day than the typical American gets in a week. Paleolithic populations were probably just as active, if not more. Some researchers have argued that the genus *Homo* evolved to run down our prey, exhausting them in hours-long chases under the hot equatorial sun. A few hunter-gatherer populations today and in the recent past do just that, taking down large antelope and other game by running them until they drop.

Regardless of whether running was the singular key adaptation that propelled *Homo* through the Pleistocene period or whether our ancestors were more generalist in their endurance behaviors, it's clear that natural selection favored a suite of traits that enhance our aerobic abilities. In traditional hunter-gatherer societies people routinely walk three or four times farther each day, and spend more of their energy working to get food, than any of the other apes. We have more fatigue-resistant "slow-twitch" muscles in our legs and greater VO_2max, a measure of aerobic capacity, than our ape relatives. Our long-legged skeletons are built to cover a lot of ground. And our hearts are built for endurance. Evolution changed the shape of the human heart to increase the volume of blood it can eject with each beat, a critical adaptation for the hours and hours of aerobic activity that hunting and gathering require each day.

What happens, though, when a body built to move spends its days slumped in an office chair or on the sofa? We've been running that experiment on ourselves for a couple of generations now. The results aren't great. . . .

PRESSURE AND TIME: HEART DISEASE AND OTHER MODERN MALADIES

The post–World War II era was a heady time in American science. Physicists at Los Alamos had split the atom, ushering in the nuclear age with all its horrors and promise. The golden age of antibiotics was also underway, with penicillin approved for clinical use during the war. Chuck Yeager broke the sound barrier, and engineers were working hard to send people into space. Across the Atlantic, James Watson and Francis Crick announced the structure of DNA, using (but not crediting) X-ray images taken by Rosalind Franklin. Psychiatrists were exploring the uses of a powerful new drug called LSD. Was there anything we clever humans couldn't do?

But a silent killer was on the loose, one that science was ill-equipped to fight. In just a few short decades, heart disease had grown to become the major cause of death in the United States. Also called cardiovascular disease, heart disease generally involves hardened or clogged arteries, especially the coronary arteries that feed the heart itself. Blocked coronary arteries cause "heart attacks" that starve the heart of oxygen, killing heart muscle cells and, if severe enough, killing the heart's owner. Hearts can also become enlarged in response to the heavy workload of chronic high blood pressure, reducing their efficiency and ability to pump blood effectively.

Unlike diseases of the past that had preyed most heavily on the poor and desperate, heart disease seemed to come for everyone. In 1950 alone, more Americans died from heart disease than had been killed in the two world wars combined. From 1923 to 1945, heart attacks and strokes (another common result of heart disease) killed five U.S. presidents, claiming Harding and Roosevelt while they were still in office. In 1955, then president Dwight

Eisenhower had a massive heart attack, nearly adding him to the list. Apart from bed rest and blood thinners, there wasn't much doctors could do for a heart attack patient or someone with heart disease. They weren't even certain why it happened at all.

The scale of the threat demanded an equally ambitious response. In 1948, researchers at the newly minted National Heart Institute launched the Framingham Heart Study, following more than 5,200 men and women in the town of Framingham, Massachusetts. It was a new approach in epidemiology: they enrolled healthy participants free of heart disease, then followed them for years to see who got sick and who didn't. Researchers could then look back at the data from earlier in the study to figure out what variables predicted a heart disease diagnosis.

The first reports pointed to two risk factors: blood pressure and cholesterol. An otherwise healthy fifty-year-old man with a systolic blood pressure (the top number of a blood pressure reading) of 180 and total cholesterol of 300 had a one-in-five chance of developing heart disease within six years. Lower the blood pressure to 130, and the risk was less than one in ten. With a cholesterol of 200 and systolic pressure of 130, the risk was less than one in twenty.

In London, another cause of heart disease was being uncovered by researcher Jerry Morris and his team in another first-of-its-kind study. Heart disease was a leading killer there, too, and Morris noticed a striking pattern: those who were physically active in their daily lives seemed to be protected from disease. In a landmark study, he compared drivers and conductors on the famous London buses. These men were similar in age, worked the same hours, spent time in the same parts of the city, and had similar incomes, but the conductors walked up and down the aisles all day collecting tickets, while the drivers sat. That small difference held big consequences. Drivers were nearly twice as likely as conductors to experience and die from a heart attack. Morris found similar results across other occupations as well. Cushy, low-activity jobs, including the coveted white-collar jobs that marked professional success in the modern era, were killing people.

Seventy-five years later those initial results hold up. High blood pressure is bad. Hearts, like any muscle, can grow in response to the extra work of pumping at a higher pressure, and enlarged hearts don't pump blood efficiently. Your vessels also have to absorb the pounding pressure wave of each heartbeat, and high pressures can take a toll, particularly on the smallest vessels, like those in your kidneys and brain and the thin coronary arteries that wrap around the heart itself. That's why high blood pressure is also a risk factor for kidney disease, and why it can lead to cerebral hemorrhage, a type of stroke in which a vessel in the brain tears and blood gushes into the skull, squishing the brain and starving neurons of oxygen.

Cholesterol is generally bad, too, particularly the "LDL" variety. Cholesterol tends to build up in artery walls, causing lumps called "plaques" that can clog the artery. When these soft plaques break apart, they can cause even more trouble. In the thin coronary vessels that feed the heart, plaques that break apart can create larger clogs as they heal over. Broken bits of plaque from other arteries can float around in the bloodstream and block small, crucial arteries in the kidneys, brain, or other organs, starving them of fresh blood and oxygen. Blocked arteries are bad, regardless of the cause. If the clog happens in your brain, we call it an "ischemic" stroke. If it happens in the thin coronary vessels that feed the heart, we call it a heart attack or, if you prefer, a "myocardial infarction."

And physical activity is *really* good for you. People who get two or three hours of exercise each week are far less likely to die from cardiovascular disease than people who don't exercise at all.

Exercise gets your heart pumping and blood rushing through your arteries, an important stimulus to the cells in the artery walls that helps to keep them elastic and feeling young. More is generally better, but the good news is that the benefits of exercise accrue rapidly. If your life is completely sedentary, adding even a little bit of movement to your day substantially lowers your risk of falling over dead from a heart attack.

What do high blood pressure, high cholesterol, and low physical activity have in common? They are all new. Just a few generations ago—the blink of an evolutionary eye—people were active every day and had blood pressures and cholesterol numbers that would make a cardiologist drool. As researchers began to collect these data in farming and hunter-gatherer populations around the world in the early 1900s, communities that had resisted industrialization and the transition to supermarkets and office work, they were surprised to find blood pressures were remarkably low and didn't increase with age in older adults. People spent the day on their feet, walking and working. Cholesterol levels were low in these populations as well, likely a consequence of their highly active lifestyles and diets low in saturated fats (we'll dive deep into diet in the next chapter).

We see this today in our work with the Hadza hunter-gatherer community. We collect blood pressures, cholesterol levels, and daily activity as part of our normal research on Hadza health and lifestyle. Men and women typically have total cholesterols around 110, with LDL levels below 70. The average blood pressure for adults over sixty is just 126/70, the same as adults in their forties. The Hadza diet varies day to day and month to month, but they don't eat nearly as much fat as Americans do. And they get over two hours of physical activity (mostly walking) every day, even into their old age.

The healthiest hearts in the world are found among the

Tsimane community, a population in the Amazon basin of rural Bolivia. Their blood pressures and cholesterol levels are similar to the Hadza, and their days are filled with physical activity. Farmed plantains are a staple of the low-fat Tsimane diet, along with a mix of wild plants, game, and fish. When researchers used CT scans to search for calcified plaques in the coronary arteries of over seven hundred Tsimane adults, they found only 3 percent with coronary artery calcification scores over 100 (the threshold score for hardened arteries). The proportion of Americans with hardened arteries was five times higher.

We need to be careful when we look to populations like the Hadza and the Tsimane as windows into the past. No population today is an ancestor, trapped in some Paleolithic time warp—we're all modern humans, the whole world 'round. But the lesson we can learn from farmers and foragers is clear: people who live off the land and get a lot of exercise maintain exceptional heart health even into old age. And we were *all* living off the land until just a couple centuries ago. Our bodies evolved over millions of years to meet the demands of a very active lifestyle. When we take that away, as we've done rapidly with industrialization, our hearts and vessels suffer. We can learn this from our neighbors like the Hadza and Tsimane, or find the same lessons on a London bus.

GENETICS AND GEOGRAPHY IN HEART DISEASE

The battle against heart disease in the United States has been a deeply unsatisfying success. Growing awareness of the major risk factors, along with the development of effective drugs to treat high blood pressure and cholesterol, have greatly reduced heart disease's toll. Public health campaigns against smoking, another major risk factor for heart disease, have helped as well: far fewer people smoke today than did a generation ago. In the 1960s, nearly 0.5 percent of the U.S. population died each year from heart disease. By 1990, that rate was cut in half. In the decades since, it's been halved again. No one can celebrate 660,000 adults dying in the U.S. last year from heart disease and strokes, but if people were dying at the same rates as 1980, that number would be over 1 million. We're making progress.

But the gains have been incredibly uneven. In the United States, Black, Hispanic, and Native American adults are much more likely to develop and die from heart disease than White men and women. Racial disparities in heart health seem to track well-established inequalities in income and access to health care, part of the larger legacy of racism in the U.S. Yet the consensus in American medicine through the 1990s and 2000s, and still heard today, is that some racial groups, particularly Black Americans, are just genetically susceptible to heart disease. Could genetic differences underlie the racial disparity in heart health?

Genetic effects certainly aren't out of the question. Work from the Framingham Heart Study, well before the modern genetics era, established that heart disease is heritable: if your parents or other blood relatives develop heart disease, you're at increased risk yourself. Work over the past few decades has clarified the role of genetics. For instance, we now know that there are several genes with alleles that increase LDL cholesterol levels, and that people who carry those alleles have a higher risk of developing heart disease. GWAS studies and other approaches have identified hundreds of other alleles that influence your risk of developing heart disease. Still, DNA isn't destiny. Twin and family studies produce values around 0.40, while GWAS studies give estimates closer to 0.10. As we saw in our discussions of IQ in the last chapter, a heritability of 0.10 means that it's nearly impossible to predict any individual person's risk of heart disease from their DNA.

Even with the higher heritability values from twin studies, there's no evidence that genetic differences can explain racial disparities in heart disease risk. Heritability studies can't tell us anything about differences between groups. And frankly, we shouldn't expect variation between populations in heart function. Hearts and vessels have the same job no matter where you live, so there haven't been the consistent, localized selection pressures needed to create local adaptations.

There *are* a handful of local adaptations that affect the blood. The *EPAS1* allele, which we discussed already, is common in Himalayan populations and protects against the overproduction of red blood cells at high altitude. *ABO* blood type alleles are another example. We each carry two copies of the *ABO* gene (as is the case for nearly all our genes), one on the chromosome 9 we inherited from mom and one on the chromosome 9 we inherited from dad. The variants, or alleles, that we have at each of those genes are either *A*, *B*, or *O*, and they determine whether our blood type is A, B, AB, or O (the "positive" or "negative" part of your blood type is determined by a different gene). Blood types and allele frequencies vary geographically. For example, *B* accounts for 20 to 30 percent of the alleles in central and western Asia, but is rare among native populations in the Americas. *A* is relatively common in Scandinavia, Australia, and some arctic North American populations, but less common in western Africa. *O* alleles are the most common worldwide, ranging from around half to nearly all of the *ABO* alleles in a population depending on the region. It's not entirely clear how this patterning of ABO blood types came about, but researchers have generally concluded that certain alleles were beneficial in particular regions, providing some immunological advantage against the local pathogens encountered.

There are also a handful of genes affecting red blood cells that have been under selection for malaria resistance. Mosquitoes, which carry the malaria parasite, are responsible for millions of deaths worldwide each year, making them easily the most dangerous animals on the planet. Malaria has been a serious threat throughout the tropics for tens of thousands of years, and only grew worse as population densities increased with farming. It's exactly the kind of strong, local, consistent environmental pressure that can lead to adaptations.

The most famous anti-malaria adaptation is the sickle cell allele, so called because it has a tendency to turn normally disk-shaped red blood cells into half-moon shapes. Like the *ABO* blood type alleles, sickle cell is a simple Mendelian trait: you can carry zero, one, or two copies of the sickle cell allele, and that determines your phenotype. One copy of the sickle cell allele confers protection against malaria infection—the malaria parasites have a hard time using red blood cells as a nursery to reproduce, which is their usual trick. Two copies typically cause sickle cell disease, in which too many of the red blood cells are half-moon shaped, causing them to clog up capillaries and do a poor job carrying oxygen. Sickle cell disease is usually fatal without modern medical intervention, which would usually mean it would be eliminated by natural selection. But the benefit for those who carry *one* copy of the sickle cell allele is so strong in malaria-infested areas that natural selection kept it in the gene pool. Adaptations rarely arise without trade-offs.

Adaptations in the blood, like sickle cell, *EPAS1*, and *ABO* alleles, are windows into our evolutionary history and the diversity that allows us to thrive anywhere on the planet. But they can't explain racial differences in heart disease. For a start, there's little evidence that any of the known blood adaptations meaningfully affect heart disease risk. Some studies suggest that having *A* or *B* blood type alleles increases heart disease risk slightly, but other studies disagree and the effect is small if it's there at all. Carrying one copy of the sickle cell allele doesn't affect your likelihood of developing heart disease or having a stroke.

More important, the geographic distribution of these alleles doesn't map onto racial groups. The frequency of the *B* allele, for instance, varies across populations within Africa, but is similar in Central Africa, Siberia, Eastern Europe, and Scandinavia. The *O* allele, which might provide some slight protection against heart disease, is found at similar rates across much of Africa, Europe, and Asia. Sickle cell allele is associated with recent African ancestry in the U.S., but there are large parts of sub-Saharan Africa where it's absent, and populations outside of Africa, in Greece and India for instance, where it's not uncommon. Races are made-up social categories, not biological groups. Genetic diversity doesn't sort neatly into racial groups because race isn't genetic.

Race can *become* biological, though, if a person's race shapes their environment. As we discussed in earlier chapters, growing up with poor nutrition affects our bodies and changes gene activity in ways that increase the likelihood of developing obesity, heart disease, and other problems. Chronic stress can alter our bodies as well. When we are upset, uncertain, or unhappy it ignites our sympathetic "fight or flight" response, raising our blood pressure. If those stresses never go away they can also lead to chronic inflammation, an overactive immune system response we'll discuss in chapter 8 that irritates blood vessels and causes them to harden.

Anyone can grow up with poor nutrition and chronic stress, but we know their impact is wildly uneven in the U.S. Black, Hispanic, Native American, and other minority communities are far more likely than White communities to experience food insecurity, and they live with the chronic stress of racism. They are also less likely to have access to health care. No wonder the people in these communities are more likely to develop heart disease and other serious medical problems.

There is a growing awareness that social conditions matter for cardiovascular health. Black Americans who experience more racism tend to have increased levels of inflammation, which we know increases disease risk. In 2023, a long-term study of forty-eight thousand Black women in the U.S. found that those reporting higher levels of racism at work, in their housing situation, and with the police were 26 percent more likely to develop heart disease. Conversely, when we examine populations outside the U.S., we don't find elevated cardiovascular disease risks among Black people. In Nigeria, average blood pressure is similar to that of White Americans. The Hadza have remarkably healthy hearts. The Tsimane, a population native to the Americas, have the healthiest hearts on the planet, in stark contrast to the high rates of high blood pressure and cardiovascular disease among Native American populations in the U.S. Racial health disparities are biological problems that arise from social inequality, not genetic differences.

PICK A CARD, ANY CARD

Life is a game of chance. The long-term odds are against you—the house always wins in the end, whether it's a mismatch disease that gets you or something else. But it would be nice to stay at the table for a long while, and to enjoy the time we're here. How do we minimize the odds of leaving early? There's a lot that's out of your control, from the alleles you carry to the color of your skin, but there's a lot you can do to stack the odds in your favor.

You can avoid smoking, for a start. You are twenty times more likely to die from lung cancer or chronic obstructive pulmonary disease (diseases of the lung like emphysema) if you smoke even a few cigarettes per day. You also double your risk of dying from heart disease and stroke. Everyone loves a cozy fire, but it turns out that turning your lungs into a chimney is really bad for you.

Reducing the stress in your life is a great idea, if you're able to do it. Chronic stress is an important risk factor for heart disease, and the more you experience, the greater your risk. But we can take steps to help ourselves. In a 2012 study with Black men and women, those randomly assigned to a meditation group reported less anger and, more important, had their odds of dying from heart attack or stroke cut in half compared to the control group who didn't meditate. It would be better, of course, if chronic stress wasn't woven into the fabric of modern life to begin with. Until that changes, we need to approach meditation, mindfulness, art, play, and anything else that lowers our stress levels as health care, not luxuries.

Staying active every day like your Paleolithic ancestors did is a great way to reduce stress and has the added advantage of keeping your vessels feeling young and elastic. You don't need to cosplay like the shirtless Paleo bros that prowl the internet. Any exercise or physical activity helps (unless you're running to the store for cigarettes). Like the London bus conductors, if we can find a way to stay on our feet in the modern world, we'll be less likely to die. Maintaining a healthy weight, something we'll discuss in detail in the next chapter, also reduces your risk of heart disease.

The grim reality, though, is that we can do all these things and *still* end up in the hospital. Chance permeates our biology, and nowhere is it a more mischievous demon than with our hearts. Nearly all of us in the industrialized world will grow old with at least some plaques in our arteries and a few hardened vessels. Every pressure wave from every heartbeat is a chance for a plaque to break apart and send a clot hurtling toward disaster, or for an artery to burst in your brain. It's a roll of the dice. Bad things happen to good people. And, perversely, fate seems to smile on plenty of folks who are openly asking for it. My lovely grandma Jean, an unrepentant smoker who enjoyed whiskey drinks with lunch, lived to be ninety-three. *See also: Keith Richards.*

To understand problems like heart disease and the importance of lifestyle we need to learn to think in shades of probability and risk rather than black-and-white certainty. Life is a gamble, so let's think like gamblers.

Imagine playing a game we'll call Queen of Hearts. The rules are simple: I hand you a well-shuffled deck of playing cards face down, and you pick the top card. If it's the Queen of Hearts, game over, you lose: you will *definitely* develop heart disease at some point within the next ten years. If it's not, I put your card back in the deck, reshuffle them thoroughly, and we play again.

Your lifestyle determines the number of times we play: healthier behaviors mean fewer rounds. The chance of drawing the Queen of Hearts in any round is about 2 percent.* If we play two rounds, the chances of drawing a Queen of Hearts in at least one of the rounds is 4 percent, in three rounds it's 6 percent, and so on. According to the American Heart Association's PREVENT risk calculator, my chances of developing heart disease in the next ten years is about 2 percent, the same as my odds of drawing the Queen of Hearts in one round. It's not a guarantee I won't get heart disease, but it doesn't keep me up at night. If I were a smoker, my risk of developing heart disease in the next ten years would be 4 percent, the same as drawing the Queen of Hearts once in two rounds. Now I'm getting nervous. . . . If my systolic blood pressure were higher, say 170 instead of 125, my odds increase again to 10 percent, equivalent to five rounds of Queen of Hearts. I don't like that at all.

---

* With a thoroughly shuffled fifty-two-card deck, your odds of pulling the Queen of Hearts off the top is 1/52, or 1.92 percent.

Notice that the odds aren't very high. I'm in my forties, before heart disease typically creeps in, and my blood pressure and cholesterol are in the healthy range. Still, smoking doubled my risk of disease. In public health jargon, my "absolute risk" increased from 2 percent to 4 percent, but my "relative risk" increased by a whopping 100 percent. Changes in relative risk are typically much larger, and sound more impressive and click-worthy, than changes in absolute risk, which are usually small. Relative risk is also an incredibly important metric in public health, since doubling the relative risk of a disease will mean twice as many people get sick. In a large population that's a big deal.

Is a 4 percent chance worth worrying about for you as an individual? Depends on your perspective, but your appetite for risk is probably lower than you think. It helps to have some context. BASE jumping, arguably the most dangerous sport in the world, involves jumping off of thousand-foot cliffs with little more than a nylon parachute in a backpack (brightly colored flying-squirrel wingsuits are optional). It seems like the sort of activity designed to kill people. In fact, BASE jumpers have an accident about every 230 jumps, a risk of 0.4 percent per jump. In five jumps, a BASE jumper incurs a 2 percent risk of an accident, the same risk as developing heart disease in ten years for a forty-five-year-old male nonsmoker. Smoking adds the equivalent of another five jumps' worth of risk. If jumping off of a thousand-foot cliff an extra five times sounds unreasonably dangerous, perhaps you ought to put the cigarettes away.

The Queen of Hearts game also shows how random chance can throw a wrench into our intuitions about risk. Even a relatively healthy person like me can get heart disease—it's not *crazy* to think the top card of the deck might be the Queen of Hearts (it's far more likely than my chances of crashing in a BASE jump). And the smoker version of me is unlikely to get sick, even if his chances are worse than mine—the odds are twenty-four to one that he'll avoid the Queen of Hearts for two rounds. If I live long enough and know enough people, I'll eventually find examples of people who got sick unexpectedly and those who surprisingly stayed healthy. That's the nature of gambling. Your lifestyle choices matter! But they aren't a guarantee.

The potential for confusion is catnip for self-help charlatans and conspiracy theory grifters. They can sell useless supplements to an anxious public and let randomness prevail. Most people taking the stuff are bound to stay healthy *and* know someone who doesn't take it and gets sick. If you paid $99.99 for a box of magic placebo pills, you're inclined to interpret that as proof you were smart and not just lucky.

Even more insidious are the folks, some of them doctors, pushing conspiracy theories about medicines shown in numerous well-designed studies to reduce heart disease risk for those with high blood pressure, cholesterol, or other risk factors. Modern blood pressure meds, for example, are known to reduce the risk of heart attack and other cardiovascular disaster in people who have high blood pressure, with minimal side effects.

Some anti-medication fearmongering comes from overinflating other risks. For example, statins are known to reduce the risk of heart attack and death from cardiovascular disease in people with elevated LDL cholesterol levels, but they also increase the risk of developing type 2 diabetes. However, the trade-off in risk is often fairly clear: one recent analysis calculated that for every one hundred people with high LDL cholesterol and high risk for cardiovascular death, statins will prevent around ten cardiovascular events like heart attack or stroke over a five-year period, while triggering diabetes in one person. Anti-med grifters also like to focus on the small change in absolute risk that many medications confer, but remember that absolute risks *always* look small. Small doesn't mean insignificant. If you can decrease your absolute risk of a catastrophic heart attack or stroke by just 0.4 percent, it's the same as sparing yourself one BASE jump.

But people don't like taking medicine in general, certainly not medicines they don't need, and we all love to hate big corporations like pharmaceutical companies. If you tell people they should stop the meds, that the whole thing is a conspiracy by doctors who want to keep their "customers" sick, you'll get plenty of skeptical people to sign on. And even though stopping those meds demonstrably increases their absolute and relative risks of heart disease, the grifters are lauded as heroes. Most of their marks don't get sick (absolute risks stay fairly low), and the ones that do are too dead to complain.

The biggest arena for all this manipulative misdirection is your dinner table. There's an entire subculture of cholesterol deniers, some with MD after their names, telling people not to worry about their cholesterol levels. Some of these same influencers push high-fat diets likely to increase your cholesterol. We know from decades of data and tons of studies that they're wrong, that living with high cholesterol (particularly LDL) increases your risk of clogged arteries and heart disease. But they continue to flourish for the same reason that people continue to jump off of cliffs with a parachute: most days, it's fine.

We've hardly touched on diet, but it's at the core of every lifestyle strategy and health guide. There's no shortage of loud debate over how to eat healthy, with the never-ending diet wars and their back-and-forth about the foods that will kill you or save your life. Can we eat eggs? Pasta? Meat? Tofu? What happens to our bodies when we do? And what happens if we eat too much? Is it true that obesity increases our risk of disease, or can we be healthy at every size? If we are what we eat, what are we meant to be?

5

GUT INSTINCT

I like to think I'm an adventurous eater, but I have strict rules against warthog soup. I didn't know this about myself until presented with the dish one hot afternoon in Deduako, a small Hadza camp tucked away in the hills above Lake Eyasi, the briny geographical anchor of their homeland. Our research team had hiked up to Deduako one afternoon for a quick census and health assessment of folks at the camp. We were greeted warmly by the residents, most of whom seemed to be taking a break from the late-day sun after a long morning of foraging. One older woman was tending an old aluminum cooking pot over a small fire. The contents looked and smelled vaguely like old roadkill stewing in a muddy puddle.

What's cooking? I inquired by way of saying hello. I half expected the answer to be something other than food. Preparing some sort of glue, perhaps.

Warthog! came the answer. And then, in a slow-motion horror sequence that I relive in my nightmares, she dipped her long-handled spoon into the mire and extended a steaming chunk in my direction. *Try some!* The smile on her face was pure grandmotherly joy.

I forced my best, most appreciative grandsonly smile back. I have eaten my share of Hadza foods, from fibrous, bland tubers to rich, spicy honey to old, undercooked zebra. I have yet to get sick from it. But a loud voice from deep within my brain was throwing an absolute tantrum about the prospect of warthog soup.

I declined as politely as I could. She stopped making eye contact.

Frankly, things could have been a lot worse. At least the warthog was cooked and, as far as I could tell, relatively fresh by Hadza standards. When a hunter gets something large, like a zebra or a giraffe, it takes several days for the camp to eat its way through every part of the animal. There's no refrigeration in Hadzaland, where the equatorial sun warms the air to oven-like temperatures every afternoon. The typical approach to storing meat is to hang pieces from tree branches and let it ripen in the heat, flies dancing all around. Sometimes the aging process starts before the meat even gets to camp, with people happily scavenging old lion kills or carrying home putrid hunks of a giraffe that was finally located days after a hunt. Rotten meat is a regular part of the Hadza menu.

They are far from alone. John Speth, an anthropologist at the University of Michigan, has done extensive research tracking down written accounts from early European explorers that document dietary customs of hunting-and-gathering cultures around the world, before they transitioned to store-bought food. The results will curl your hair. All around the world, people *love* rotten meat.

Consider, as just one lurid example, this account of fine dining from one of Robert Peary's expeditions across Greenland in the late 1800s. Ikwa, an Eskimo guide, had found a seal cached by hunters some time before and was excited to share the delicacy with Peary's crew. As he brought the delicacy back to camp, one team member recalled, "the air became filled with the most horrible stench it has ever been my misfortune to endure. . . . [Ikwa was] carrying upon his back an immense seal, which had every appearance of having been buried at least two years. Great fat maggots dropped from it at every step." Peary's crew was horrified and refused to come anywhere near it. Ikwa was hurt, telling the crew "the more decayed the seal the finer the eating, and he could not understand why we should object."

An enduring love of putrid meat was not confined to the Arctic. Dietary accounts from every continent and climate are alive with maggots, worms, and the soft, smelly flesh of decaying animals. Many groups preferred rotten meat to fresh, often burying the animals they hunted for weeks or months to get the flesh nice and ripe. Rotten fish, bison, turtles, deer, and eggs were regular items on the menu the world over. The more it stank, the better.

Diet gurus and internet influencers often tout "natural" diets, urging us to eat like our ancestors. It's persuasive marketing (no one wants to eat an *unnatural* diet), and the problems wrought by our modern food systems are very real, but the definition of ancestral diets typically feels like a caricature. Paleo bros lurk behind every kombucha stand, tan and shirtless, telling you to eat more organ meat and give up carbs. They take tourist trips to African tribes, run barefoot through forests and supermarkets, and rant about the dangers of seed oils. Theirs is a sanitized, Instagram-ready version of the past that's long on machismo and short on evidence, with a professed delight in eating testicles that seems nothing short of Freudian. They never mention the maggots.

Work by Speth and others, including research my collaborators and I have done with living hunter-gatherer communities like the Hadza, provides a more complete picture of hunter-gatherer diets and the diversity of foods and foodways our ancestors likely followed. Diets are diverse and always have been, shifting across time and geography as local communities capitalized on the resources around them. There was never any singular or uniform ancestral diet. Lots of foods prized by traditional groups are disgusting to our modern sensibilities, putrefied and crawling. All of this raises the question of what and how we ought to eat. What would a natural, ancestral human diet look like, and would we actually want to eat it?

THE BEST WAY TO UNDERSTAND HOW FOOD AFFECTS YOUR BODY IS to follow a meal. In honor of my wife, a Philadelphia native and die-hard Eagles fan, let's imagine tearing into a steaming Philly cheesesteak. Ground beef, spray cheese, and some sautéed peppers and onions, nestled in a white-bread bun, beckon to you lasciviously from a paper plate. *Is the plate reused from a previous diner? Is the guy in line behind you Hall of Fame quarterback Ron Jaworski? Why isn't anyone behind the counter wearing a hairnet?* You think of Ikwa and push those thoughts away. You're in the cheesesteak's world now.

Digestion begins as soon as you start chewing. The evolutionary history of a species is written in its teeth—the flat, grinding molars of a grass eater or the slicing blades of a carnivore. Early human ancestors (the genus *Australopithecus*) had big molars (cheek teeth) with thick enamel, adapted to a diet that leaned heavily into hard and fibrous plant foods. With hunting and gathering, tooth size in early *Homo* got smaller. Meat and organs are easier to chew, and adding them to the diet meant the energy and nutrients used to build and operate a big chewing apparatus could be better spent elsewhere. Alleles for smaller teeth were favored. Tooth shape stayed roughly the same, reflecting mixed diets of plant and animal foods. Our teeth are broadly similar to a pig's or a bear's, good for breaking down a wide range of foods rather than specialized for any one thing.

Saliva helps lubricate that delicious bolus of cheesesteak as you chew, though it's hardly necessary with all the squeezy cheese and burger juice. Produced in glands that sit toward the back of your cheek and tingle when you eat something sour, saliva also carries an enzyme called amylase. Digestive enzymes are the molecules responsible for breaking food down into simple, usable nutrients. Amylase chops up starches, which are just long chains of the simple sugar glucose. The bun of the cheesesteak is full of starch, as are many foods made of grains and other plant parts.

Not all animals produce salivary amylase. Like large molars, it's an evolutionary case of "use it or lose it." The amylase genes we carry reflect the importance of root vegetables, grains, and other plant foods in our evolutionary past. Fossilized plaque stuck to the teeth of Paleolithic humans typically contains grains and starch remnants as well, underlining the long history and dietary significance of starch and reminding us all to brush and floss.

Once your bite of heaven is sufficiently mashed into a pulp, you swallow it, the muscles of your throat and esophagus pushing the mush down to your stomach. The human stomach is incredibly acidic, with a pH of just 1.5, *much* more caustic than other primates, or even carnivores or plant eaters, for that matter. The only animals with such acidic stomachs are vultures and other scavengers. Stomach acid helps kill bacteria and other nasties in our food, and our high acidity is likely an adaptation to handle rotten food. That's probably why communities like Ikwa's, who made a habit of eating putrefied foods, never seemed to get sick, much to the amazement of European explorers. The hydrochloric acid in your stomach also helps to produce an enzyme called pepsin, which digests proteins like those in the burger and cheese.

After an hour or two of churning in the acidic soup of your stomach, the slurry that was once a cheesesteak is squeezed into the small intestine. Two ducts near the beginning of the small intestine squirt enzymes and other goodies into the mix. The duct from your pancreas, a small, jalapeño-shaped organ resting below your stomach, adds bicarbonate to neutralize the hydrochloric acid along with a long list of enzymes to digest carbohydrates (starches and sugars), proteins, and fats. The duct from your gallbladder, a thumb-sized greenish pouch sandwiched between your liver and intestine, adds bile. Bile acts like soap when you're scrubbing a greasy pan. It breaks fats (including oils) into microdroplets, providing lots of surface area for enzymes to digest them.

Over 90 percent of all the nutrients in your food are fully digested in the small intestine and absorbed into your body for use (the rest go down the toilet). There are three basic nutrient categories, called "macronutrients": carbohydrates, fats, and proteins. Each is broken down into its component parts. Carbohydrates, whether they are starches or sugars, are digested down to simple sugars (monosaccharides), mainly glucose. Fats are digested down to fatty acids (saturated, monounsaturated, polyunsaturated, or trans fats, depending on the details of their chemical structure). Proteins are broken into amino acids. These basic building blocks—fatty acids, amino acids, and monosaccharides—pass through the intestinal wall and into the bloodstream.* Vitamins and minerals do as well.

All populations today have the same set of digestive enzymes. Diets are so diverse and change so quickly that there's been little opportunity to develop the consistent, localized selection pressures needed for natural selection to produce local adaptations.

---

* Fatty acids take a detour through the lymph vessels on the way to the bloodstream. We'll discuss lymph vessels in chapter 9.

That said, there have been a couple interesting cases of selection to enhance enzyme activity. In populations that rely heavily on starchy foods like tubers and rice, people tend to have more copies of the gene that makes salivary amylase,* presumably under selection to digest starches more effectively. And in a couple of early farming populations that relied heavily on milk from goats and cattle, "lactase persistence" evolved, in which the gene that produces the enzyme to digest lactose, the main sugar in milk, remains active into adulthood. All mammals can digest lactose as milk-drinking infants, but the gene that produces the critical enzyme, lactase, typically deactivates after weaning. Lactase persistence evolved independently in northern Europe and eastern Africa, presumably because it promoted survival and reproduction by enabling adults in those cultures to extract more energy from the milk and cheese they depended on.

Still, neither the salivary amylase copy variation nor the lactase persistence alleles seem to have strong effects on people's ability to eat different foods. People in meat-heavy cultures still produce plenty of amylase to digest starch. And lactase persistence alleles aren't required to enjoy dairy. Many people without them appear to digest milk and cheese just fine. A recent study of over three hundred thousand adults in the United Kingdom found that more than 90 percent of adults without lactase persistence alleles regularly drank milk, the same percentage as adults with the alleles.

Anything left from your meal after all this digestion and absorption is eventually squeezed into the large intestine. There, it meets a rich ecosystem of trillions of bacteria, a hot and steamy

---

* One common type of genetic mutation is repeated segments of DNA (GACGAT becomes GACGCGCGCGAT, for example). "Gene duplication" occurs when an entire gene is repeated. As a result, individuals can differ in the number of copies they carry for a particular gene. More copies can often mean more protein is produced, and thus more starch-digesting enzyme in the case of salivary amylase.

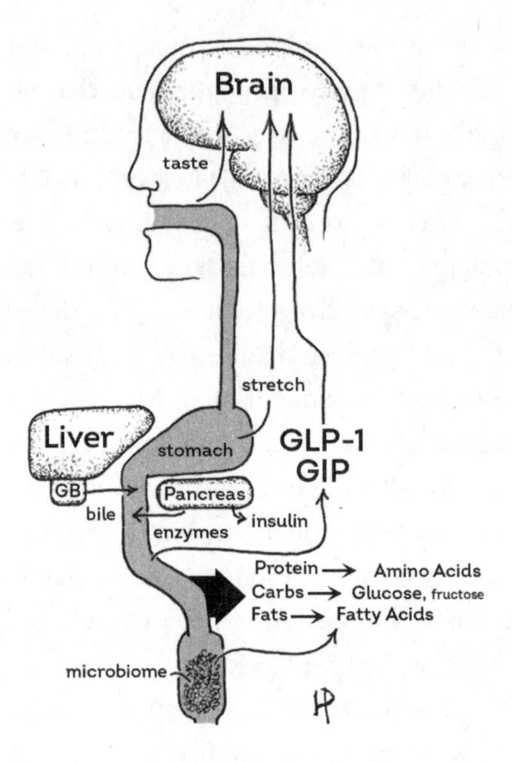

Figure 5.1 Overview of digestion. Food enters the stomach and then the intestines, where bile from the gallbladder (GB) and enzymes from the pancreas digest proteins, carbs, and fats. The pancreas will also produce insulin. In the small intestine, proteins are digested into amino acids, carbohydrates into glucose and fructose, and fats into fatty acids. These nutrients are absorbed through the intestinal wall and end up in the bloodstream. Digestion of fiber in the large intestine by the microbiome also produces fatty acids. Eating and digestion send signals to the brain, including the taste of the food, stretch sensation from the stomach, and incretin hormones (GLP-1, GIP, and others) from the small intestine. These signals are integrated in the brain, affecting feelings of hunger and satiety, and ultimately determining how much you eat.

inner swamp called your microbiome. The bacteria feast on fiber from your diet, which the enzymes in your small intestine can't digest. Bacterial digestion produces fatty acids that we're able to absorb into the bloodstream. The large intestine also pulls water out of the soupy muck it receives from the small intestine, recycling that water back into the bloodstream. The result is that familiar, compact brown conglomerate of undigested food, cells sloughed off of the intestinal wall, and bacteria that we call poo.

Across primates, species with more easily digestible diets tend to have smaller digestive tracts—the evolutionary "use it or lose it" principle in action. Stretched end to end, stomach to anus, the average human intestinal tract is twenty-seven feet (eight meters) long. That may sound impressive, but it's marginally shorter than those of orangutans and chimpanzees in proportion to body size. Our large intestines, in particular, have shortened a bit. As with our smaller teeth, hunting and gathering and the move to include more meat in the diet seems to have made an overbuilt digestive tract unnecessary. Hunter-gatherer communities in warm parts of the world also eat a lot of honey, another high-calorie food. Cooking turbocharged the diet further, increasing digestibility. Potatoes, for example, provide double the number of calories when they're cooked versus raw.

The consequence of all these changes is an incredibly energy-dense diet, with nearly twice the calories per gram of a typical primate diet. As a result, humans burn through *more* calories each day even though we eat a *smaller* volume of food. Those extra calories helped to fuel the evolutionary increase in brain size over the past 2 million years.

The evolutionary history written in our teeth and guts reflects the energy-rich mix of cooked plants and animal foods that have characterized the diet since the Stone Age. Humans thrive on a broad range of diets, from meat-heavy to vegetarian. We see this dietary diversity among hunter-gatherer populations today and in the recent past. In warm climates, where there's a mix of plant and animal species available, we see a highly diverse mix of diets. The *average* diet is a rough balance of plant and animal foods, but the *variation* in diets is the real story. Populations like the Hadza eat a lot of different foods, and the mix changes over time. Meat might make up a big part of the diet one month but be scarce the next. Some weeks, honey accounts for nearly half their calories. On balance, the groups we know the most about, like the Hadza, usually eat more carbohydrates and less fat than people in the U.S., in sharp contrast to meat-centric "ancestral" diets touted online.

What we *don't* see in our digestive system is evidence for specialization. A big part of being the most successful and prolific primate species is our ability to eat a broad range of foods, making the most of whatever the local environment has to offer. We eat everything.

PUTTING NUTRIENTS TO WORK

Once nutrients are absorbed into the bloodstream, the real metabolic fun begins. "Metabolism" is a broad term for the work that your 37 trillion cells do all day as they chug along like microscopic factories. Glucose, fatty acids, and amino acids provide the fuel and the raw materials. All that work takes energy, which we measure in calories, or joules. Your "metabolic rate" is the speed at which you burn those calories.

Amino acids are the building blocks for nearly everything that cells make, from enzymes to new tissue to a range of hormones. Building proteins from amino acids is what genes are all about. The DNA sequence in a gene acts as a template for lining up amino acids and assembling them into long chains called proteins (that's why slight differences in the DNA sequence, variants that we call alleles, can lead to different proteins produced). All proteins are made from just twenty-one different types of amino acid. For about half of them, our cells can convert one type of amino acid to another. The others, called "essential" amino acids, we need to consume in our diets.

Glucose and fatty acids are mostly used as fuel, providing the energy needed to power cellular construction projects and to move our bodies around. These molecules, made of carbon, oxygen, and hydrogen atoms stuck together like Lego bricks, are broken apart in the cells. The energy released as they're broken apart gets stored in molecules called ATP, which our cells use like rechargeable batteries. Most ATP production happens inside chambers called mitochondria, which, if you remember anything from high school biology, you'll know as the "powerhouse of the cell." The chemistry of ATP production in the mitochondria requires oxygen, which is why *you* need oxygen: no oxygen, no life, for animals like us.

The waste from all of this energy production, the scraps left over after glucose and fatty acids have been dismantled like a stolen car, is CO_2. It's dumped into the bloodstream and exhaled (see chapter 4). That means you'll eventually exhale your cheesesteak, just as you exhale every bit of food you eat and burn off. Meanwhile, the oxygen molecules you inhale will pick up hydrogen atoms in the mitochondria and form water. It's the everyday alchemy of life, more than a billion years old. Food becomes air, air becomes water.*

There are additional complications to this basic scheme. Fructose and galactose, the other, less common monosaccharides in the diet, are first converted to glucose for use by the cells. Amino acids can be converted to glucose and utilized to make ATP. Some sugars and fatty acids are used to build structures like DNA and cell membranes. Fatty acids can also be converted into molecules called ketones, which are then used to make ATP.

Ketones become an important fuel when glucose is scarce. They can be burned by nearly any cell in the body, but are

---

* It's also the mirror image of what plants do. Photosynthesis turns CO_2 into glucose and splits water molecules apart to form oxygen. Animals and plant metabolism balance one another on a global scale, each breathing in what the other breathes out.

particularly important for the brain, which can *only* use glucose or ketones (not fatty acids). People who are starving, relying on stored fat for energy, will produce and burn ketones; it's an evolved strategy that we see in a number of species, including our orangutan cousins. People on "keto" diets, which are very low in carbohydrates (less than fifty grams per day), and therefore low on glucose, produce and burn ketones as well.

Of course, you don't burn all the energy from your meal right away. A foot-long Philly cheesesteak is well over one thousand kilocalories,* enough energy to power your protein robot for hours. Letting all the glucose and fatty acids absorbed from the meal float around in your blood until they were finally burned would cause serious problems. Instead, all the glucose and fatty acids we don't use immediately need to be stored. That's where insulin comes in.

Insulin is a hormone produced by the pancreas, with wide-ranging effects throughout the body. Insulin production is triggered by rising glucose levels in the blood in the wake of a meal that includes carbohydrates. Fatty acids and amino acids also increase insulin production, as do hormones like GLP-1 and GIP, which are produced by the intestines in response to a meal. Insulin helps to convert glucose into stuff called "glycogen" for storage in the liver and muscles. But the body's ability to store glycogen is limited, and once capacity is reached any additional glucose is stored as fat. Fatty acids are also stored as fat. There's no real upper bound for the amount of fat a body can carry (the heaviest person on record weighed 1,400 pounds), but there are good and bad places to park it. Once the usual spaces are full, the body

* For silly historical reasons, the "calories" on our food labels and diet plans are actually *kilocalories*. One *calorie* is tiny, too small to be useful when we talk about food—you eat millions of them each day. Saying "calories" when you mean "kilocalories" is like saying "meters" when you mean "kilometers," or "yards" when you mean "miles." For all that is good and decent, *please* stop.

starts packing fat in and around the organs, which leads to a whole host of health problems like the aptly named fatty liver disease.

Eventually, with the excess glucose stored away, the energy demands of your cells will begin to draw down the amount of glucose circulating in the blood. When blood glucose levels fall, the pancreas stops producing insulin and starts producing the hormone glucagon. Glucagon is the yin to insulin's yang, turning glycogen back into glucose and breaking down fat for fuel. The system is beautifully balanced, at least for the young and healthy. Insulin builds our energy stores, glucagon draws them down. Blood glucose never strays beyond a narrow range. Fat is a fuel tank, a buffer between meals, kept comfortably between empty and full.

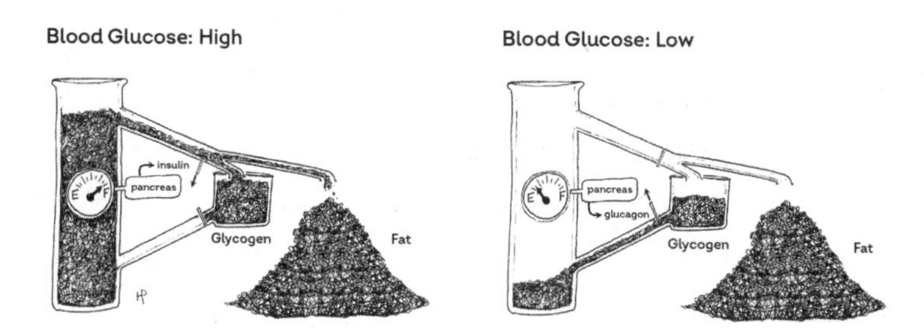

Figure 5.2 Blood glucose regulation. The pancreas senses blood glucose (blood sugar) levels. When blood glucose is high, the pancreas produces insulin, which channels glucose to glycogen and fat. When blood glucose is low, the pancreas produces glucagon, which converts glycogen to glucose. Glucagon also promotes burning fat for energy. In type 1 diabetes, the pancreas cannot produce insulin. In type 2 diabetes, insulin is produced but cells don't respond to it. With both types of diabetes, blood glucose climbs to dangerous levels.

For some, the system falls apart because critical pieces go missing. For example, in type 1 diabetes, also called juvenile-onset diabetes, the insulin-producing beta cells in the pancreas die, leaving the person with an impaired ability to make the critical hormone. Glucose storage is compromised, as is the ability of cells to pull glucose in through their membranes. With no place to go, blood glucose levels skyrocket, damaging capillaries and causing all sorts of problems.

But for many more, the system seems to come unbalanced in subtle and slippery ways. We store more fat than we burn. Not a lot, and not every day, but slowly the *drip, drip, drip* of extra calories adds up on the bathroom scale. The typical American adult gains a pound or two per year—just a *smidge* out of balance, but enough to cause trouble in the long run.

Every ounce of you was built from the food you ate, carbs and proteins and fats that you ingested, absorbed into your bloodstream, and transformed into a human. Eventually they'll be broken down and burned for energy or discarded as waste. We use calories (or joules) to track the flow of nutrients into and out of your body the way that we use gallons (or liters) to measure the flow of water into and out of a lake. If the calories you eat match the calories you burn, your weight won't change. Get out of calorie balance, and your weight will change accordingly. Your weight really is as simple as *calories in, calories out.* Anyone who tells you differently is either misinformed or selling you a fad diet.

But the simple math of calories and weight doesn't mean the obesity crisis is simple to fix, or that all foods are the same, or that maintaining a healthy weight is easy. Millions of research hours and billions of dollars have been spent trying to figure out whether the modern epidemic of obesity is primarily a "calories in" problem caused by our diet or a "calories out" problem in the ways we burn energy. A solution to the crisis remains elusive, but we've learned a lot about how our evolved genes and modern environments interact to push many of us toward obesity. Let's start with "calories out," your metabolic rate.

URINE FOR A SURPRISE

It's never not weird to ask someone to pee in a cup. Stranger still to be the person asked to do the peeing. Yet here we were, Onawasi and I, on a bright morning in Hadzaland, a small plastic Dixie cup between us. An older man in his sixties, Onawasi had never been in a doctor's office to play out this scene, but the absurdity of it seemed to transcend time and place. Hardened by a lifetime spent earning his dinner on the dry savanna, but warm and quick with a smile, he had a reputation of being a bit of a rascal. He took the cup with a smirk and tromped off behind an acacia tree.

My colleagues and I were working with the Hadza that summer, in 2010, to measure daily energy expenditures, the total number of calories burned each day. The urine samples were an integral part. Participants like Onawasi drank a small dose of water enriched in rare isotopes of hydrogen and oxygen, and we tracked the flow of those isotopes through the body over the course of a week, using urine samples taken every couple of days. The speed with which the isotopes were flushed from the body told us precisely how much CO_2 each person was producing, and CO_2 production, as we saw earlier, is directly tied to the production and burning of energy. This technique, called the "doubly labeled water" method, is completely safe and highly accurate, the gold standard for measuring the total calories burned per day. It's used in studies all over the world, but we were going to be the first to use it in a hunter-gatherer population.

Making a living off the land is a lot of work, and communities like the Hadza get more physical activity in a day than most Americans get in a week. Many in public health assumed, quite sensibly, that the reduced physical activity typical in industrialized societies was a major cause of the obesity crisis. Less energy spent on activity, by their logic, meant fewer calories burned each day and more calories stored as fat. It made sense, but no one had done the measurements. We didn't *really* know how daily energy expenditures in the U.S. and other industrialized countries today compared to those in the deep past. Obviously we couldn't go back in time to measure daily energy expenditures in our ancestors, but we could measure the metabolism of people living hunter-gatherer lifestyles today. We were in Hadzaland to get these crucial data.

What happened next turned everything I thought I understood about metabolism on its head. The Universe had a trick up its sleeve. Call it Onawasi's revenge.

When we analyzed all those urine samples back in the U.S., at one of the best doubly labeled water labs in the world, the numbers that came back seemed impossible. Daily energy expenditures among the Hadza men and women in our sample were no different from those of U.S. adults and other industrialized populations. Hadza adults were much more active each day, yet they burned the same number of calories as sedentary adults in the developed world.

It turned out this wasn't a fluke. Work done by my colleague Amy Luke, measuring daily energy expenditures among women in rural Nigeria, had found a similar result, with the same daily expenditures as women in Chicago. A large meta-analysis out of Luke's lab, examining studies from populations around the globe, had found no relationship between economic development and daily expenditure. And follow-up work from my lab and collaborators found no difference in daily expenditure between farmers, foragers, and pastoralists, like the Tsimane, Shuar, and Daasanach, and populations in industrialized communities. Populations following hunter-gatherer and farming lifestyles similar to those in our collective past, with much more physical activity each day, somehow burned the same number of calories each day as people working desk jobs in developed countries.

We're still working to understand how the body manages this feat, which I explore at length in the book *Burn* and we'll discuss in more detail in the next chapter. Somehow, in people with more active lifestyles, the body adjusts the way it spends its energy. More energy is burned by their muscles, but less is spent elsewhere, on physiological tasks like reproduction or immune function. These adjustments are likely a big reason that exercise is so good for us. But regardless of how it's done, the implications for the modern obesity crisis are clear. Today's weight problems can't be due to lower daily energy expenditures from our less active, modern lifestyles. As far as we can tell, daily energy expenditures today are the same as they were in the past, before industrialization, when we were hunting, gathering, and farming.

Could *individual* differences in energy expenditure explain recent changes in obesity? Sure, industrialized populations today might have the same energy expenditure as those in the past, but within any population there will be variation in metabolic rate, people with "fast" metabolisms that burn more energy or "slow" metabolisms that burn less. Perhaps people with slow metabolisms, burning fewer calories each day than typical for their age and body size, are more prone to gaining weight in our modern environments. It's an appealing explanation, and one you hear a lot—people often chalk up their problems with weight to a slow metabolism. But here again, the data throw cold water on our intuitions.

Metabolic rates certainly do vary among individuals. The biggest factor is your body size. The larger you are, the more cells you're made of, and the more energy they'll consume. Some organs are greedier than others. Brains, livers, and kidneys burn more energy than other organs. Fat burns the least. So it matters how large you are and how much of you is fat. Bigger people, and those who carry more lean mass and less fat, burn more energy.

For example, a typical man in an industrialized population like the U.S., at 180 pounds (82 kg) and 24 percent body fat, burns 3,060 kilocalories (12.8 MJ) per day, while a typical woman, at 165 pounds (75 kg) and 37 percent body fat, burns 2,365 kilocalories (9.9 MJ) per day. The sex difference in energy expenditure is entirely due to the difference in size and body fat percentage. Similarly, when we compare expenditures across populations we always account for body size and composition. Hadza men and women actually burn *less* energy per day than adults in the U.S., but the difference disappears when you account for the difference in body size. Hadza adults tend to weigh considerably less.

Age also affects your metabolic rate. With colleagues from metabolism labs around the globe, I recently analyzed daily energy expenditure measurements for over six thousand people, from babies just eight days old to men and women in their nineties, something no study had been able to do before. We tested whether age affected metabolic rate once we controlled for the effects of body size. The results were eye-opening. Young children burn less energy each day than adults do because they're smaller. But for their size, young children burn far *more* energy than adults—their cells are working overtime, growing and developing their organs and body systems. Metabolic rate peaks around your first birthday and then slowly declines, reaching adult levels in your early twenties. Daily energy expenditure is remarkably stable through adulthood, into your late fifties. We don't see any changes with menopause, for example, or the slowdown with middle age that people tend to blame for their expanding waistlines. Then, around age sixty, metabolism starts to decline. Cells slow down, becoming less active, reflecting the decline in our bodies with age.

Even after we account for body size, fat percentage, and age, there's still a lot of variation in the energy burned each day. As we see with the Hadza study, very little of this variation can be attributed to differences in daily physical activity. Two adults with the exact same body weight, fat percentage, and lifestyle can easily differ by five hundred kilocalories per day or even more. We don't have a great understanding of why some people burn more and others less. Some of this difference is genetic. The heritability of metabolic rate hasn't been studied very extensively, but family studies give estimates of around 0.40 (GWAS estimates would likely be lower). It's not clear how these genes affect metabolic rate, but there are a range of possibilities. For example, these alleles could affect the production of thyroid hormone, which stimulates cell activity throughout the body.

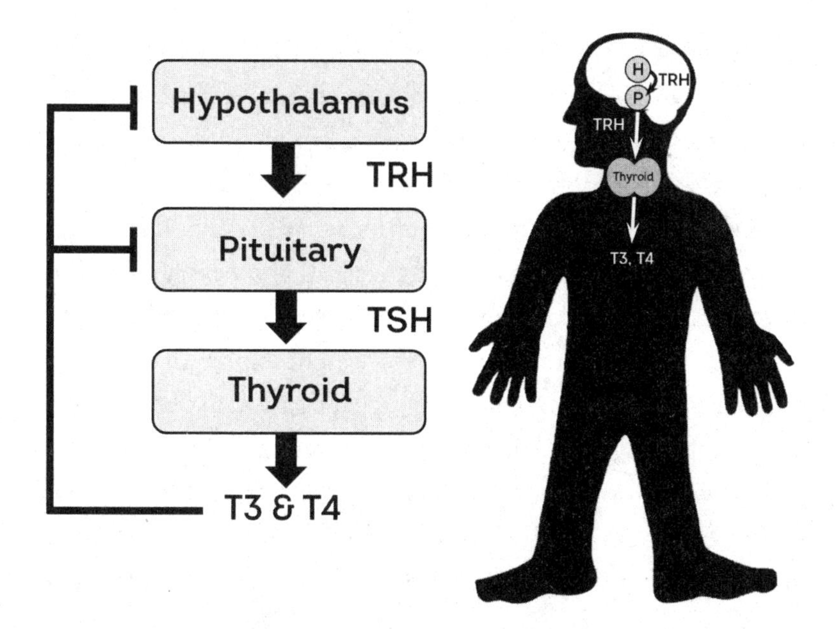

Figure 5.3 Thyroid hormone is produced by the thyroid gland, a butterfly-shaped organ in your neck. The hypothalamus (H) produces TRH, stimulating the pituitary gland (P) to produce TSH, which in turn stimulates the thyroid to produce the two major forms of thyroid hormone, T3 and T4. You need iodine to make T3 and T4, making it an essential nutrient. Today, table salt and other foods have added iodine for just that reason.

Environment plays a big role in shaping metabolic rates, but we still don't have a good handle on the details. Given that childhood environments can impact adult body weight (chapter 1), it's likely that nutrition during development influences adult metabolism, but those effects haven't been studied. Stress and anxiety could be important factors. A study measuring energy expenditure at rest, lying quietly on an exam table, found that male college students with high anxiety levels burned about 6 percent more energy just lying still than those with low anxiety levels. Weight loss can also impact metabolism. Many people experience a slower metabolism when they lose a lot of weight, and this effect can last for years.

Whatever causes the variation in metabolic rate we see among individuals, it's stable. If you have a fast metabolism when we measure you today, burning more energy each day than we'd expect for your body weight and fat percentage, you'll likely have a fast metabolism when we measure you again in a month or a year or a decade from now. If you're slow, you'll stay slow.

But guess what? It doesn't matter. People with a fast metabolism aren't protected from gaining weight, and people with a slow metabolism aren't doomed. Men and women with obesity have the same metabolic rates as normal-weight adults. In fact, in raw numbers, not accounting for body size, people with obesity tend to burn *more* energy each day because they are larger.

The weight-loss world is full of products and programs to boost your metabolism. Some might be helpful (most of us could use more exercise in our lives). Others might be harmful, like the unregulated miracle supplements hawked by influencers all over the internet. All of them are empty promises. Exercise is fantastic for your health, but its impact on the calories you burn each day is negligible. Supplements and shakes won't boost your metabolism, either. And even if they did work to boost your metabolism, they'd have little impact on your weight. Obesity isn't a "calories out" problem. We aren't growing fat as a society, or as individuals, because we're not burning enough energy. Obesity is a "calories in" problem. We're eating too much. The question is, *why*?

ET TU, CRÈME BRÛLÉE? WHY THERE'S TOO MUCH OF TOO MANY OF US

The foods we eat have changed radically in the past century, by nearly every nutritional metric. Modern diets pack more calories into each bite. They're higher in saturated fat, salt, and sugar, and undoubtedly more delicious. Fiber and protein are scarce. Strange new chemicals have appeared. Each change presents our body with a mismatch between the types of foods we evolved to eat and the strange new products we consume today. Obesity is a problem created by this evolutionary mismatch, born from these dietary changes—but which ones? It's like an Agatha Christie murder mystery, the suspects standing around a fresh corpse as a thunderstorm crackles outside. Any of these changes could be the culprit, and they've all been accused.

A popular and provocative idea, developed in the early 2000s by Gary Taubes, David Ludwig, and others, blames sugars and other carbohydrates for the obesity crisis. Modern carb-heavy diets, they argue, cause an overproduction of insulin that, in turn, sweeps a large portion of circulating glucose and fatty acids into fat cells, leaving your blood low on fuel. Your brain is tricked into thinking you're starving even though you've just eaten. Subconsciously, your brain tries to fix the problem by pushing you to eat more while simultaneously lowering your energy expenditure. Carbs and insulin push you to overconsume and pack away too much fat.

This carbohydrate-insulin model has led to a popular revolt

against counting calories, with folks arguing that calories don't matter, that it's all about the carbs. But like a lot of revolutionary ideas, it hasn't held up to careful testing. There's no doubt that a lot of people lose weight when they cut sugars and other carbohydrates out of their diet. But if carbs were uniquely dangerous, then low-carbohydrate approaches should be superior to low-fat diets for weight loss. They aren't. In head-to-head trials, people lose weight on low-fat diets, like Mediterranean and Ornish, the same as they do on keto diets and other low-carb approaches.* You can randomly assign people to a range of diets, and the only predictor of weight-loss success is whether they follow their diet.

Low-carb diets work because they cut calories, and they're popular because some people find them relatively easy and painless to stick with. They certainly aren't magic, nor are they the only diets that work. But if it's not all about the carbs, what else could be fueling the obesity crisis?

Another explanation, developed by Stephen Simpson and David Raubenheimer, is that modern diets are too low in protein. As they discuss in their book *Eat Like the Animals*, many species, including some of our primate relatives, seem to target a particular amount of protein each day, eating a variety of foods until they hit their mark. If foods are low in protein, these animals end up consuming more carbs and fats, and therefore more calories, to meet their protein goal. There's evidence that humans follow a similar pattern, eating until they reach some internal target for protein. And there's data from work in the 1990s that people find higher-protein foods more satiating, feeling fuller on fewer calories.

---

* People even lose weight on *all*-carb diets, like the potato diet, in which you eat nothing but boiled potatoes (no butter, salt, or other flavoring). Mark Haub, a nutrition professor at Kansas State University, lost twenty-seven pounds in two months on an all-junk-food diet of snack cakes and chips. It's the calories, not the carbs.

Other researchers have focused on energy density. Modern foods often pack more energy into each bite than our diets would have in the past, making it easy to consume an enormous amount of calories without feeling like you've eaten that much food. Plant foods you find in the wild, like the tubers that Hadza folks eat almost daily, pack a lot of fiber, which fills you up without adding to your calorie intake. Wild game, like antelope and giraffe, is also leaner, carrying much less fat (and calories) than the farm-raised animals at your local butcher.

Carbs, protein, fat, fiber . . . you can find convincing arguments for any of them (and others), each with their own social media tribes and influencers shouting from the ramparts. But rather than a classic murder mystery, where we're looking for a single actor, it seems increasingly apparent that the better analogy is Shakespeare's *Julius Caesar*. Our waistlines are attacked daily by a mob of modern dietary assailants. It's death by a thousand cuts.

Kevin Hall, a researcher at the U.S. National Institutes of Health, has shown just how problematic our modern foods can be. His group fed study participants two different menus, one made entirely of ultra-processed foods (prepackaged dinners, meat products, boxed cereals, and the like) and another made of minimally processed foods (roasted chicken, steamed broccoli, that sort of thing). Each participant ate both diets, each for two weeks, and were instructed to eat as much as they wanted. The diets were identical in the amount of carbs, proteins, and fat and similar in energy density. Despite these similarities, participants ate five hundred kilocalories per day *more* when they were given the ultra-processed diet, and they gained a pound per week. On the minimally processed diet, they lost weight.

Like pornography, there's no simple definition of "ultra-processed foods" that everyone will agree on, but you know it when you see it.* Premade and prepackaged, long ingredient lists full of substances that would make a chemist blush, added sugars and oils, mind-bendingly delicious. Ultra-processed foods combine the worst of our modern dietary problems. They are energy dense and high in added carbs (mostly sugars) and fats (mostly oils), making them easy to overeat. They're low in protein and fiber, and therefore less filling. Most are engineered to light up your brain's reward circuits with a mix of flavors that has never existed in earth's history—flavors so rewarding that they're addictive. And they are taking over. More than half of the calories consumed in rich countries like the U.S. come from ultra-processed foods. They pack our supermarket shelves and fast-food restaurants. No wonder we have an obesity problem.

If ultra-processed foods are so pervasive and problematic, why isn't *everyone* obese? The standard story is a morality tale—people with obesity just don't have the willpower to eat healthy. Decades of research have demonstrated that view is neither true nor helpful. Instead, body weight is a clear example of genes and environment colliding to shape our bodies.

Family studies of obesity report heritability estimates of around 0.40, while GWAS studies give estimates of about 0.20. Like height, body weight is influenced by many genes (over nine hundred), each with a tiny effect. Studies that have tracked down these genes have shown that they are active in the brain, not in other organs. In the past, when ultra-processed foods weren't around, none of this genetic diversity would have caused trouble. The alleles were probably neutral, with no impact on survival or reproduction. (It's even possible that alleles encouraging more consumption could have been beneficial, pushing our ancestors to

* The most common classification system, NOVA, is often used in research studies, but isn't universally accepted.

pack away a few more calories as a hedge against lean times.) Today, in our modern environments, these alleles cause problems.

Obesity is a problem of eating too many calories, and the system that regulates what we eat resides in our brains. The gene variants you inherit influence the way you're wired when you come into the world, which in turn shapes the way your brain responds to modern foods and whether you'll struggle with your weight.

DO WE WEIGH WEIGHT TOO HEAVILY?

Obesity is unfair. It is nearly impossible to avoid ultra-processed foods, and the problems they cause fall unevenly across society. We can't change our genes or the circumstances we're born into. Once we're overweight or obese, it's very hard to change. Most diets fail. The stigma around body fat can be cruel and damaging to mental health. Perhaps we'd be better off, as folks in the Health at Every Size movement argue, if we stopped worrying about our weight and focused instead on other measures of health.

BMI (body mass index), a ratio of weight to height that's used to compare body size and fatness among people of different stature, has come under particular scrutiny. Like weight, BMI is a rough measure, unable to distinguish whether a person is carrying a lot of muscle or fat. It's also a continuous measure that's chopped into categories—underweight, normal, overweight, or obese—to label people as healthy or unhealthy. Those cutoffs are arbitrary and can perpetuate racial disparities in health care. As we discussed in chapter 2, body size and proportion are heritable, and some populations are inherently thinner or stockier than others, which leads to differences in the amount of fat carried for a given BMI. The cutoffs for one population might not be appropriate for another. For example, Black adults in the U.S. tend to carry more lean muscle and less fat at a given BMI than White adults do, and therefore the standard BMI cutoffs, which were developed using White populations, will tend to unfairly label more Black adults as overweight or obese. Conversely, Asian adults tend to carry more body fat at a given BMI, and the standard cutoffs can underestimate their risk. An Asian adult with a BMI of 28 has a much higher risk of developing obesity-related disease than a White or Black American at the same BMI. Many, including some in the Health at Every Size movement, have called for BMI to be abandoned as an outdated and racist metric.

It's shameful that people with obesity face discrimination and stigma, and there's no doubt that we as a society need to do better. But I don't think that we can afford to ignore body weight or BMI in public health or in our own lives. The stubborn fact remains that carrying a lot of body fat is a serious risk factor for disease. BMI provides a useful, if imperfect, metric of body fat without the need for expensive equipment or complicated measurements. We'd be foolish to throw it away.

Having a BMI above 30 (the typical cutoff for "obese") puts you at increased risk for a range of health problems. As we discussed with smoking or high blood pressure in the last chapter, obesity doesn't guarantee you'll get sick, but it increases your odds. In a recent study of over 3.6 million adults in the United Kingdom, people with obesity were more likely to die from heart disease, some cancers (particularly of the liver, kidneys, or uterus), respiratory failure, diabetes, and liver disease. They were also more vulnerable to infections, something we saw with the COVID pandemic. (Thinner isn't always better. Being "underweight," with a BMI less than 18.5, also increases your health risks.)

Diabetes is a particular concern. In type 2 diabetes (also called "adult onset"), your cells become insensitive to insulin—they don't take in glucose the way they normally would when insulin is in the bloodstream. With nowhere for the glucose to go, blood sugar levels reach dangerous heights, causing damage throughout the body that can eventually kill you. BMI strongly predicts your risk of developing and dying from type 2 diabetes. In one recent study, men and women with a BMI over 30 were six times more likely to develop the disease. By comparison, carrying a high number of alleles associated with developing diabetes and having an unhealthy lifestyle (e.g., not exercising) were much less predictive. Obesity, measured by that rough metric BMI, was a much better indicator of diabetes risk than either genetics or lifestyle.

There's plenty of room for improvement in BMI as a metric for obesity. Population-specific BMI cutoffs could be useful, for a start, and we're seeing that approach emerge in public health. We'd be further along if we moved away from a race-based concept of human variation and toward a more population-based approach. Races are not biologically or genetically coherent categories, they are social constructs. The BMI cutoffs for Black people with recent ancestry from West Africa are probably inappropriate for groups like the Daasanach, whom we met in chapter 2, where people tend to carry alleles that confer a tall, thin build. Similarly, "Asian" is a nearly meaningless category, as it encompasses 60 percent of all the people on the planet and a large number of diverse populations.

Perhaps one day, with bioelectrical impedance scales and other tech becoming more widespread, we'll have accurate measures of body fat—the real variable of interest in obesity—in our homes, and we'll be able to move beyond BMI. Hopefully we'll have more effective approaches to prevent or reverse obesity as well, and less stigma around body weight. Those are all important goals to work toward. In the meantime, we can't let the problems of blunt measures like weight and BMI, or the challenges of managing our weight, keep us from addressing the problem of obesity. The health risks are too important to ignore.

FOOD AND DRUG ADMINISTRATION

The obesity crisis is a problem of our own making. In building our industrialized worlds, we've created an evolutionary mismatch between our ancient DNA and modern environments. The alleles we all carry worked perfectly well to keep us at a healthy weight for millennia. Now, exposed to strange, new, delicious foods that we've engineered ourselves, many of us find our genes are no longer up to the task. It happened fast. Alleles that didn't cause any trouble in our grandparents' generation now push many of us to overeat.

We can't change our DNA, and so our only solution is to reshape our environments, to reimagine them again. In the long term, as a society, we can change the way foods are produced, to eliminate the harmful ultra-processed foods that surround us today. We can make the old-fashioned foods that our grandparents and great-grandparents ate cheaper and more readily available. Perhaps we can even engineer delicious and shelf-stable ultra-processed foods that promote health.

For now, we can work to eliminate the foods that make us fat from our own personal environments. We would all do well to keep ultra-processed foods out of our shopping carts, refrigerators, and cupboards. That's easier if you're lucky enough to live in a neighborhood with good supermarkets. Likewise, if you grew up in a family that understood the importance of nutrition, had the money and time to shop for good food, and created a culture around healthy eating, you've got a great head start. We need to tackle economic and racial health disparities by making healthy foods and nutrition education available to everyone.

Environmental influences on obesity aren't limited to food, though. Other factors in our day-to-day lives can push us to overeat. Stress can push us to overeat, the dopamine reward of a snack or a meal providing a temporary emotional reprieve. The chronic stresses of poverty and racism amplify the problem, and may help to explain why obesity is more prevalent in poor and minority communities. Sleep is important, too. In a recent study, overweight men and women were coached to improve their sleep habits, mainly by removing television and other screen time from their pre-bed routines. They got more sleep, ate better, and lost weight.

And as my beloved grandmother used to tell me (paraphrasing), if diet and lifestyle don't work, there's always drugs. The history of obesity treatment in the U.S. is littered with failed therapies and snake oil scams, but a recently developed class of medicines has shown incredible promise. These drugs mimic two hormones, GLP-1 and GIP, that the small intestine produces in response to a meal, telling the brain that you've eaten. The synthetic versions of those hormones, sold under brand names like Ozempic, Wegovy, and Mounjaro, seem to have a similar effect, making people feel full and uninterested in eating. In clinical trials, adults with obesity lose an average of 15 to 20 percent of their body weight on these drugs and show improvement in blood pressure and other health measures as well. Previously, the only treatment for obesity with such a large and reliable impact was bariatric surgery.

The response to these drugs has been interesting to watch. With all the public hand-wringing about the perils of obesity over the past few decades, you'd expect these new drugs to be hailed as a miracle cure. Instead, influencers, lifestyle coaches, and even some doctors are hot and bothered, expressing everything from concern to outrage. It's true that there are some known side effects, mostly nausea, that need to be taken seriously. And it's fair to wonder whether these drugs will become the first and only line of recourse for people struggling with their weight. People tend to return to their old weight when they stop taking these medicines, and we should try diet and lifestyle changes *before* signing someone up for a lifelong prescription. The drugs are expensive as well, at least for now, threatening to deepen the inequities in healthcare access.

But most of the negativity around these drugs sounds like moral panic. It is deeply ingrained in our culture that health reflects good character and virtuous living. Disease is seen as evidence of personal failing. People with obesity are viewed as flawed and undisciplined. From that perspective, losing weight is *supposed* to be difficult and painful, a proportionate penance for your sins. You should be grinding it out in the gym, red-faced and sweating, the regret for your poor choices dripping from your brow. Drugs that allow you to lose weight without suffering seem like cheating.

It's an old-fashioned, Old Testament view of the world that ought to be relinquished to the dustbin of history. Yes, our choices matter and we need to take personal responsibility for our behaviors, including our diets. But that doesn't mean we'll all be successful, or that extra pounds signal some wider moral decay. Some of us have a harder job than others because of the environments we live in, the alleles we've been dealt, or both. If we find ourselves at an unhealthy weight and decide to turn things around, we should be encouraged and given every tool available. Diet and lifestyle are the right place to start, but they aren't meant to be punishments. Getting healthier shouldn't require suffering.

PLEASE DON'T MAKE ME EAT WARTHOG SOUP

Obesity is fundamentally a diet problem, so what should we eat? Paleo? Mediterranean? Vegetarian? Keto? Intermittent fasting? Is it time we all listened to Ikwa and embraced rotten meat and maggots? (Still waiting for the Paleo movement to get *real* and start promoting that. . . .)

Pay no attention to the evangelists who claim there's only one right answer. Humans are evolved to thrive on a wide range of diets. Any of them might work for you, and they all work the same way, by providing a healthy number of calories while keeping you satisfied. The one that works best for you will depend on your personal history with food and the ways those experiences have wired your brain. You'll know your diet is a good fit when it gets you moving toward a healthy weight and doesn't feel like a miserable slog.

Diet has effects far beyond our waistlines, and we should pay attention to other aspects of health as well. The trick is that isolating the effects of particular foods or nutrients is incredibly difficult. For one thing, any particular food is likely to play a small part in the variety of foods you eat over the course of weeks and months. Also, anything you remove from your diet will be replaced by something else, making it hard to know if the changes you experience come from the subtraction of the old food or the addition of the new. As a result, foods that look good in one study can sometimes look unhealthy in another, adding to the confusion around diet.

What we can say with confidence is that your healthiest diet has nothing to do with your blood type. None of that is real. If you follow a type B diet and are loving it, *great*! You have randomly stumbled upon a diet that works, and I'm happy for you. But there's no evidence that your blood type has any effect on the diet that's healthiest for you. Same for zodiac signs.

The evidence for personalized diets based on your DNA is equally thin at the moment. It's possible that we'll develop meaningful dietary recommendations based on your unique genome—people are working on it nonstop, and it's plausible that the alleles you carry affect the way you digest, metabolize, and respond to foods in ways that affect your health. We have yet to see that connection bear out, however. As with other complex traits, it's likely that the predictive power of genetics will remain small, because environment has such a large effect. Time will tell.

For now, we're stuck with general guidelines that we can tweak to fit our tastes. Protein and fiber are good for you. Added sugars and fats (including oils) are bad. Carbs aren't the devil, particularly when they're part of fiber- and vitamin-rich veggies and fruits. Fats aren't inherently bad, but you should lean more toward unsaturated fats (poly- or monounsaturated) and less toward saturated fats, which increase cholesterol levels and your risk for heart disease, as we discussed in the last chapter. If you want to get even more picky, you can opt for foods with more omega-3 fatty acids, which may reduce inflammation a bit, rather than omega-6 fatty acids, which are thought to be pro-inflammatory. Avoid trans fats, like "partially hydrogenated" oils, altogether. You should probably eat more beans, which are high in protein and fiber and low in sugars and fats. Cured meats are consistently associated with a range of health problems, from stomach cancer to lung failure, so eat them sparingly. Avoiding ultra-processed foods is a good way to incorporate a lot of these healthy guidelines.

One advantage to trying a mixed diet with a balance of plant and animal foods is that it's generally easier to get all the vitamins, minerals, and other nutrients your body needs. Fruits and vegetables are easy sources of fiber (aim for at least twenty-five grams per day) and the water-soluble vitamins C, B, and folic acid. Meat, fish, and dairy are easy sources of iron and calcium as well as the fat-soluble vitamins A, D, E, and K. The more extreme or restrictive the diet, the more you'll want to be careful that you're getting the full complement of nutrients and not overdoing it on any one thing. Vegan diets, for example, can be a great option, but you'll want to make sure you're getting enough iron, calcium, and protein (around fifty grams/day for women, sixty for men as a rough guide). Carnivore and keto diets are typically low in fiber and can also be high in saturated fat.

The good news is that none of this requires us to eat putrid meat. Ancestral diets are a nice idea in the abstract, and eating fresh, unprocessed foods that your grandparents would recognize is generally a good idea. But we don't need to pretend to be hunter-gatherers, and it wouldn't work anyway. It's hard to imagine a profitable business based around warthog soup, for a start, and the foods in your supermarket—even the ones in the vegetable aisle and butcher's counter—have been changed through thousands of years of farming to be vastly different from their wild ancestors. We can't go back. The only solution is forward. We were clever enough to design the complex food environment we live in today. Hopefully we're smart enough to make it better.

While we're at it, we should engineer our environments to promote physical activity as well. Exercise might not do much to boost your metabolism or reduce your weight, but it's still incredibly important for health. Just as we evolved to eat a diverse range of foods, our Paleolithic ancestors had diverse lifestyles attuned to their local environments. But the common thread across all cultures prior to industrialization was physical activity. Your protein robot is built to move.

6

MUSCLE AND BONE

If aliens from the far reaches of the galaxy ever visit our planet, I hope they get here during the Olympics. It would be a nice crash course on the diversity of our species. Massive powerlifters, petite gymnasts, rail-thin marathoners, and muscular sprinters. The triumph of the human spirit. Experimental drug use and geopolitics. There's no stronger display of the interplay between genes, environment, and chance in shaping our bodies and our lives.

Millions of kids dream of Olympic glory, but vanishingly few achieve it. You can be sure that the ones who do carry more than the average number of alleles promoting mental toughness, competitive drive, and athletic talent; that they've practiced thousands of hours in the best nutritional and training environments to develop their craft; and they've had the good luck of being born into the right circumstances, playing the right opponents, being healthy at the right time, to make it onto their national squad.

Lots of variables go into the preparation and specialization of elite athletes, but fundamental to it all is *muscle.* Aside from fat, no other organ in your body has the capacity to shape-shift so dramatically. One athlete might be 110 pounds of leathery endurance, while another with the same frame could bulge with 250 pounds of explosive power. The difference in physique lies in how they've grown their muscle. We can see and feel these differences in ourselves as well when we fall in love with a new sport or exercise routine, or fall into sedentary habits and feel ourselves growing soft. Muscle activity and exercise impact every other system in our body, from the ways we respond to stress to our reproductive function. The muscle you carry also predicts how well you'll age, and whether you'll stay sharp and hardy into your pickleball years.

Exercise and environment aren't the only factors shaping our muscles and athletic prowess. Genetics seems to play a role as well. Our intuition that athletic talent is unevenly distributed among us rubs up against the uncomfortable observation that some groups of people dominate particular sports. It seems that Kenyans and Ethiopians win all the marathons, Black athletes from the Caribbean excel in sprinting, and White Americans are overrepresented on the podium in swimming. Is this evidence of real, racial differences in muscle physiology and athletic ability, as some like to claim, or is something else at play? What separates the champions on the podium from the weekend warriors sweating in obscurity?

GETTING IN ON THE ACTIN

The building blocks of muscle are unfathomably ancient. Two long, thin molecules—actin and myosin—form the fundamental, functional unit of muscle cells in all animals, from humans to hummingbirds, starfish to fireflies. These molecules initially evolved to move things around inside the cells of our single-celled ancestors, and they're still used for that today. More than half a billion years ago, as multicellular life evolved, cell groups specialized to particular tasks and became organs. Muscles specialized to change shape, to contract and relax, using the magic of actin and myosin.

Muscle cells, which we can also call muscle fibers, are long and thin. Like thick yarn or a steel cable, if you zoom in you'll see they are actually many thinner fibers, called "myofibrils," bundled together. Each myofibril, in turn, is a chain of thousands of microscopic units called "sarcomeres" joined end to end to end. The bundle of myofibrils is wrapped in a thin film, the cell membrane, that keeps them bathed in water and other molecules, like ATP, that they need to function.

Each sarcomere, or link in the chain, is made of overlapping actin and myosin fibers. To picture it, imagine taking the heads of two toothbrushes, one with fine, thin bristles (actin filaments) and the other with thick bristles (myosin filaments). Face those two brushes toward each other, bristles to bristles, and push them together. Hundreds of thick myosin bristles slide between hundreds of thin actin bristles as the toothbrush heads move closer together.

Now let's focus on just one pair of bristles, one thick myosin fiber pressed alongside a thin actin fiber. Myosin filaments aren't smooth like a toothbrush bristle. Instead, they are covered in thousands of microscopic projections. You can picture each projection as an arm* that reaches forward, grabs the actin filament, pulls it back, and lets go. Each myosin arm will repeat this cycle again and again, pulling the actin filament along, powered by ATP, the energy-carrying molecule we met in the last chapter. The arms are all oriented the same direction, and all go through the

---

* Somewhat confusingly for this analogy, these projections are called myosin "heads," not arms, by physiologists. So, if you want, you can picture an arm with a head on the end, grabbing on to the actin filament with its teeth.

same cycle, each in a fraction of a second. Each cycle requires one molecule of ATP.

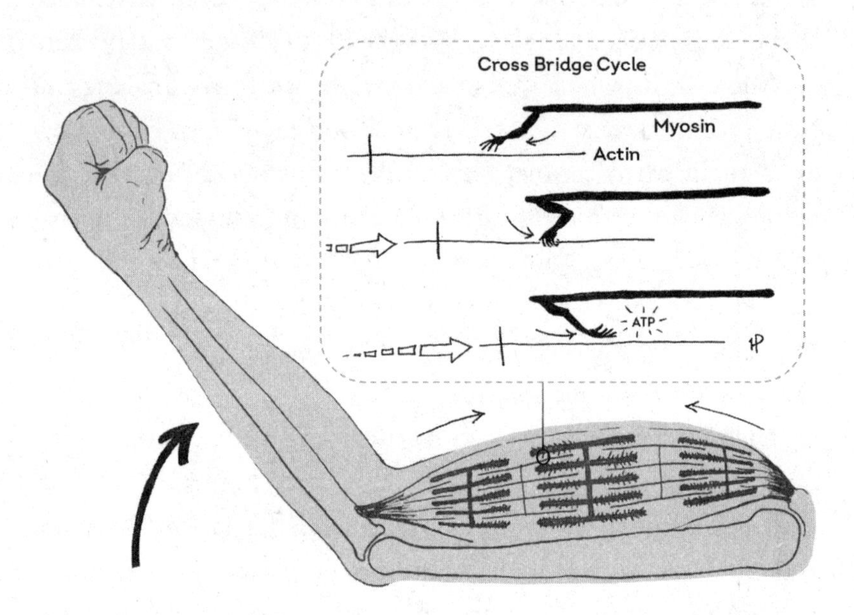

Figure 6.1 The cross-bridge cycle. Muscles are made from myosin and actin units called sarcomeres. Three sarcomeres are illustrated in the muscle here (real sarcomeres are microscopic, and a real muscle is hundreds of thousands of sarcomeres long). Where myosin and actin filaments overlap, thousands of projections from the myosin filament grab the actin filament and pull, over and over again, for as long as the muscle is stimulated. Each grab-pull-release cycle is a cross-bridge cycle, requiring one molecule of ATP. Cross-bridge cycling causes the muscle to shorten. In the case of the biceps muscle illustrated here, muscle shortening causes the elbow to flex.

Like a queue of burly sailors working together to pull in a long anchor rope, thousands of myosin arms pulling on the actin filament cause it to slide past the myosin filament. That same pulling and sliding occurs at every pair of myosin and actin filaments. Returning to our toothbrush analogy, the thick-bristled myosin brush will pull the fine-bristled actin brush deeper into itself. The toothbrush heads will move closer together. That's a muscle contraction. The muscle fiber shortens because every sarcomere (or pair of toothbrush heads) shortens. As you pick up a book or a cup of coffee and flex your elbow, that's the sarcomeres in the biceps muscle in your upper arm shortening and pulling the radius bone in your forearm toward the humerus bone in your upper arm.

The cycling of myosin grabbing and pulling on the actin filaments only happens when the muscle is stimulated by the electrical charge of a nerve signal. The action potential from the neuron hits the muscle cell membrane at a special synapse called a motor end plate. This stimulation causes a muscle action potential (similar to the action potential we learned about in axons) to sweep across the surface of the muscle cell, releasing calcium from internal reservoirs. The calcium allows the myosin to grab the actin filaments and pull.*

Once the nerve stimulation stops, the muscle action potentials stop, and the calcium molecules within the muscle cell are sucked back into their reservoirs. The myosin lets go of the actin. The filaments are free to slide past each other, and the muscle goes limp. You relax your biceps muscle and let your arm hang at your side. Your elbow straightens as the actin and myosin fibers in the sarcomeres slide past one another until just the ends overlap.

Your muscles, the lumps of meat that bulge beneath your skin, are made of thousands or even millions of muscle fibers wrapped together in a clear, tough sheath called "fascia." At their ends, the fibers join into ropelike tendons that connect muscles to bones. Your Achilles tendon, for example, connects your calf muscle to your heel bone, or calcaneus. (Ligaments are similar ropelike structures, but they connect bones to bones, with no muscles in between.) A bigger, stronger muscle has more muscle fibers packed

---

* Because muscle action potentials are electrical, muscles will also contract when shocked by an external electrical charge. That's why electric shock causes muscles to seize.

into it. More fibers, working in parallel, mean more myosin arms pulling on actin filaments to shorten the muscle, and therefore more force.

Muscles can only exert force when they shorten; they can only pull. Your protein robot body is a complex marionette, with muscles pulling their tendinous strings to move your bones. Every movement is produced by contracting one or more of our muscles. With more than six hundred muscles in your body, there are more possible combinations of muscle contraction than there are grains of sand on earth. It's no small marvel that we learn to use this complex machine, or that all the subtle, powerful, soulful, coordinated motions our bodies can perform are made possible by complex signaling from our brains.

GROWING MUSCLES AND MINDS

Muscles adapt to the ways we use them, the classic example of environmental influence on our bodies. The body is evolved to spend its energy wisely, and it only builds the muscle it needs and can afford. If we're sedentary or starving, our bodies break muscle down to burn as fuel. When we exercise and have adequate nutrition, the body responds by building muscles bigger. Exercise causes tiny rips in the muscle fibers from the wear and tear of repeated contractions. As we recover afterward, particularly when we sleep, "stem cells" in the muscle repair the damage, and grow additional fibers in preparation for next time. Stem cells are special cells in each of our organs, left over from our earliest embryonic days, that direct the growth of new tissue to repair or replace damaged cells.

Muscles need to be broken down by exercise in order to grow back bigger. That's why exercise programs designed to increase muscle size and strength emphasize powerful movements with heavy resistance. The trick is to create as much damage as possible without overdoing it and causing large tears in the muscle or tendon. Muscle strains occur when large portions of the muscle fibers break apart, and the tears are so painful and debilitating that you need to rest the muscle completely while it heals.

Exercise also affects the type of muscle we grow. There are two basic types of muscle, categorized by the way they get their ATP. Type I muscles rely on ATP made by mitochondria, using oxygen. Oxygen-based, or "aerobic" metabolism provides a nearly endless source of energy, but it takes time to go through all the steps needed to make ATP. That's why these "slow-twitch" muscles are great for endurance events like running a mile or a marathon, but not ideal for short bursts of pure power.

Type II muscles are "fast-twitch," producing more force, more quickly, and relying less on mitochondria for ATP. Much of your type II muscle doesn't carry mitochondria at all, relying instead on "anaerobic" production of ATP that doesn't require oxygen. Type II muscles are perfect for short bursts like a sprint or a dead lift, but they fatigue quickly.

You're more familiar with these types of muscle than you might realize. The dark meat in a chicken or a turkey is type I muscle. It's the muscle in the legs and thighs, and it's dark because it contains myoglobin, a cousin of hemoglobin that helps pull oxygen from the blood into the muscle. Turkeys and chickens spend most of their day standing and walking, and the type I dark meat in their legs provides the power for that aerobic activity without fatiguing. Those muscles also hold fat, which they use to make ATP via the mitochondria, which is why dark meat is juicier. Breast meat is anaerobic type II muscle, used for short bursts to fly and escape danger. It's white because it doesn't rely on oxygen and therefore doesn't have any myoglobin. Fast-twitch white meat holds very little fat as well, because it lacks the mitochondria needed to burn fat for ATP.

Athletes who train for endurance events build more type I slow-twitch muscle. Aerobic exercise also causes the mitochondria in your type I muscle to proliferate, improving the muscle's ability to use oxygen to make ATP. More mitochondria also improves an athlete's VO_2max, the maximum rate that the body can consume oxygen to power their muscles and their movement. Athletes who train for strength, in explosive events like jumping, throwing, lifting, or sprinting, build more type II fast-twitch muscle.

The diversity in body shape and ability that we see among Olympic athletes reflects a lifetime of training calibrated to the demands of their particular sport. Jumpers and powerlifters both need explosive strength. Jumpers also need to minimize body mass so they can jump higher or farther, whereas powerlifters don't pay a performance penalty for carrying extra pounds. Both end up with lots of type II muscle, but with vastly different body shapes.

We see a similar trade-off among runners. Sprinters are all power, and they need to accelerate their bodies off the starting blocks. They can't be *too* heavy, but more muscle is generally better. Distance runners need muscle, too, but the demands are different. They confront the fatigue of lifting their bodies against gravity with each step—tens of thousands of steps in a marathon. As a result, sprinters tend to be larger and packed with type II fast-twitch muscle, while distance runners are thinner and full of type I slow-twitch muscle. Decathletes, who compete in ten events from throwing shot puts to sprinting to distance running, need to decide if they want to specialize in power or endurance. Among world-class decathletes, those who do well in sprinting or throwing do worse in the 1,500-meter distance race, and vice versa.

Training also shapes the brain. The more you use your muscles, especially at high intensity, the better your brain and nervous system get at stimulating the muscle. Your body learns to activate a larger proportion of the muscle, sending more nerve signals to more motor end plates. This "neural adaptation" makes you stronger even before your muscles grow.

The brain also regulates fatigue and endurance. More type I muscle, more mitochondria, and better heart and lung function will all improve your stamina, but the feeling of complete collapse, when you can't go another step, is generated by your brain. It's an evolved survival mechanism to prevent you from pushing so hard that you damage yourself. Endurance training helps the brain grow accustomed to the signals of impending doom, giving you more time before you give in to the voice in your head imploring you to stop.

WE USED TO THINK THAT OUR MUSCLES ACTED LIKE SIMPLE ENGINES—step on the gas and watch them go. Other than the demands for ATP and oxygen, and the impact on our hearts and lungs, we didn't think that exercising muscle had a big effect on other systems in our body. Research over the past couple of decades has told a very different story.

Muscles produce hundreds of signaling molecules when we exercise, and those molecules seem to influence nearly every aspect of our body. We are still working to understand all the signals and connections, but here's what we know. People who exercise have less inflammation, apparently because the immune system responds to increased muscle work by dampening its hair-trigger response to perceived threats. Similarly, people who exercise have a blunted response to stress, their adrenaline and cortisol levels rising less abruptly and returning to normal more quickly than in people who don't exercise. Reproductive hormones like testosterone and progesterone are moderated, too, maintained at healthier, lower levels among people who exercise compared to those who are sedentary. Exercise promotes brain health as well, leading to the production of hormones in the brain that encourage neuron growth. Everywhere we look, exercising muscles affect how our bodies function. They are the rhythm section for your body's physiological orchestra.

When muscles go quiet, the orchestra falls apart. This can happen when we don't exercise enough, but it can also develop from the ways we rest. In the comfy confines of our offices and living rooms, it's tempting to let ourselves sit motionless and limp. This inactivity causes trouble. When we look at a population like the Hadza, people typically squat or kneel when they rest, and the muscles in their core and legs remain engaged. That low-level activity requires energy, and the muscles slurp up lipids (fats) from the bloodstream for fuel. In our industrialized environments, we spend hours in the suspended animation of our overstuffed recliners and ergonomic office chairs. Our flaccid muscles don't burn any energy at all, and the lipids build up in our blood. Needless to say, all the usual signals from exercising muscles are absent as well. As a result, the uniquely sedentary nature of industrialized life has become a major risk factor for heart disease and other health problems.

The wide-ranging effects that muscles have throughout your body help to explain why exercise is incredibly important for health but not particularly helpful for managing your weight. Exercise changes the way your body spends its calories, pushing it to spend less on inflammation, stress response, sky-high reproductive hormone levels, and other tasks. These changes are usually beneficial, and they also save energy. Less inflammation, for example, means reduced activity by your immune system. Lower epinephrine levels mean less cellular activity. We've even found evidence that exercise impacts the production of thyroid hormone, a major signaling molecule with effects on metabolism throughout the body.

Taken to the extreme, this downregulation can cause trouble. Elite athletes who maintain extraordinary exercise workloads can develop overtraining syndrome (also called "relative energy deficiency in sport," or RED-S). They spend so much energy each day to train that they don't have enough left to fuel their bodies' other tasks, like reproduction, immune function, or muscle repair. Colds and other infections last longer. Chronic injuries won't heal. Reproductive hormones crash, along with libido. Women stop getting their periods. Many of these issues can be resolved through diet—making sure to eat enough calories to power everything. But there are limits to how much energy the body can process per day. For athletes pushing the limits of what the body can handle, too much exercise pulls too many calories away from other, vital tasks.

Most of us, though, are in no danger of overdoing it, and could use more exercise in our lives to better regulate our body and keep us healthier. Just don't expect exercise to help much with your weight.

The dampening of activity in other systems seems to be why exercise doesn't increase our total daily energy expenditure the way we'd expect, a phenomenon I discuss in detail in my book *Burn*. Active populations like the Hadza burn the same amount of energy each day as sedentary Americans, and my colleagues and I think that's because the Hadza spend less energy on stress response, chronic inflammation, and the like. The flip side is that trying to increase your daily energy expenditure with exercise is pretty ineffective. In exercise intervention studies, we find that after a few months people's bodies adjust to the new level of physical activity, burning nearly the same number of calories they did before they started the exercise program. It's hard to budge your daily calorie burn very far or for very long, which makes it hard to lose weight with exercise. Exercise will keep you alive, but it won't make you thin.

Our bodies are evolved to move, and we're healthier when we stay in motion. So how much exercise should we try to get each day? Once again, groups like the Hadza offer a helpful example. Men and women in farming and foraging communities get around fifteen thousand steps per day. If we can emulate that, the statistics are on our side. As we saw with the London bus drivers we met in chapter 4, you don't need to pose as a hunter-gatherer to stay active. People getting over ten thousand steps per day, or more than 150 minutes of exercise per week, are much less likely to develop heart disease, diabetes, and other mismatch diseases.

Exercise also helps you age better. For reasons that still aren't entirely clear, adults past the age of around sixty tend to lose muscle. In some, it can start even earlier. Fighting to maintain your muscle helps keep the Reaper at bay and helps to keep you mentally sharp as you age. Older folks who become frail and weak are more likely to grow ill, develop dementia, and die earlier than those who keep up their strength.

STRENGTH THAT COMES FROM WITHIN

Much like the Olympics, the game of life provides its own series of physical tests. Lions sprint after antelope, colobus monkeys jump from tree to tree in the high canopy, arctic terns fly twelve-thousand-mile marathons between the poles chasing summer. As with decathletes, it's impossible to be good at everything, and rarely a winning strategy to try. Species tend to specialize. The plastic kinds of changes in muscle that we see with exercise help to keep animals fit for their particular events. But over long, evolutionary timescales, natural selection does much of the heavy lifting. Alleles to build type I muscles are favored when life requires endurance, type II muscles when it pays to sprint. The shapes of the bones change, too, optimizing mechanical advantage.

We can see these forces at work today in our own skeletons. Unlike other apes, we're bipedal, walking on our two hind limbs. Our pelvis, leg bones, and feet have evolved to make that happen, with skeletal evolution for bipedalism evident in the earliest members of our lineage. At around 4 million years ago, these species still had a grasping big toe, long arms, and long fingers, for climbing trees. But by 2 million years ago, body proportions and the shape of the foot and pelvis were essentially the same as yours today.

Dan Lieberman, an evolutionary anthropologist at Harvard and my PhD adviser, has argued that our modern body proportions evolved as early members of our genus, like *Homo erectus*, adapted to run down big game like antelope and zebra, pushing prey to exhaustion by relentlessly jogging after them under the hot equatorial sun. That might sound implausible as a way to make a living, but people do it. There are ethnographic accounts of hunter-gatherer communities, like the San in southern Africa, hunting like this. You can try it yourself at the next Man Against Horse marathon in Prescott, Arizona, where every year since 1983 people have been trying to outrun large herbivores for sport. Often, they succeed, with plenty of runners outpacing the slower horses. In 2016, Dan (never one to shrink from his convictions) ran the twenty-five-mile version of the race, beating all but thirteen of his equine competitors. In 2019, a human named Nick Coury won the whole thing, racing the mountainous fifty-mile course in a brisk six hours fourteen minutes and besting the fastest horse.

Dan and others have argued that endurance running was the key adaptation that put our lineage on the path toward big-brained, modern humanity. Others have pushed back (scientists love a good argument), pointing out that there are lots of hunter-gatherer communities, like the Hadza, who hardly run at all, and that running per se isn't necessary for evolution to favor endurance. Hunting and gathering is hard work, and most of it, even for groups like the San, is walking, digging, chopping, and other non-running tasks. But regardless of whether running was the centerpiece of our evolutionary strategy or just one part of our evolved tool kit, it's clear that humans are prodigious endurance athletes. Our aerobic abilities far exceed any other primate, and most other species on land. Your body is built to move.

It isn't just our skeleton that adapted for endurance; our muscles changed, too. Leg muscles grew larger to accommodate the extra burden of bipedalism. The proportion of type I fatigue-resistant fibers increased as well. In chimpanzees, gorillas, and most other primates, more than half of the muscle in their legs is fast-twitch type II. That's great for powering up a tree, leaping between branches, or sprinting from a leopard, but not good for endurance. In humans, most of our leg muscle is fatigue-resistant, slow-twitch type I. The aerobic demands of hunting and gathering, which more than tripled the mileage covered each day compared to our ape cousins, led to natural selection for more fatigue-resistant type II muscle in our legs. These endurance adaptations echo those in the heart that were mentioned in chapter 4. In the game of life, we specialized for distance.

We all share a human genome that builds an endurance-adapted human body rather than a chimpanzee-like, power-adapted body. But that doesn't mean we come into the world with equal athletic potential. The genes you carry influence ("determine" would be too strong a word) your strength, speed, and endurance.

The most obvious, in terms of their effects, are the genes that control the development of our reproductive systems. As we'll see in the next chapter, all mammalian embryos, including humans, are programmed by default to develop as females. But, if the embryo carries a Y chromosome with a functioning *SRY* gene, the proteins produced set into motion a cascade of events that cause the reproductive anatomy to develop into the male form, with testes instead of ovaries. The testes, in turn, produce testosterone, a hormone that promotes the growth of muscle and bone. Females produce testosterone, too, but at much lower quantities. As a result, biological males (those with XY chromosomes and typical development) tend to grow larger and put on more muscle than biological females.

As someone who grew up in a rambunctious household as the lone brother to three sisters, I can tell you firsthand that there's plenty of overlap in strength and size between males and females, particularly through the childhood and adolescent years. Even among adults, there are plenty of females who are bigger and stronger than males. Training, as well as the many other genes that affect body size (chapter 2), has a big impact on the size and strength of your muscles.

But the effects of testosterone are clear as well. When we compare athletic performance in boys and girls from childhood through their mid-teens, we see the male advantage in sprinting, jumping, and other sports appear at puberty. Prior to that, there's no clear advantage to either sex. After just a few years bathing in the sky-high hormone concentrations of puberty, teenage boys pull away from their female peers. As adults, the bell-shaped distributions of strength and speed for males overlap considerably—there are plenty of females who outperform males. But the average and top male athletes perform about 10 to 30 percent better in strength- and stamina-based events than their female counterparts. We can see this, for example, in the distribution of running speeds for the Boston Marathon. In 2017, the average running speed for men in Boston was 7.1 miles per hour, which was 10 percent faster than the average women's speed of 6.5 mph. The men's winner that year, Geoffrey Kiriu, ran at an average pace of 12.1 miles per hour (4:57 minutes per mile), 9 percent faster than his fellow Kenyan Edna Kiplagat, the top women's finisher, who averaged 11.1 miles per hour (5:25 minutes per mile).

More recently, studies investigating the impact of testosterone on muscle have shown how its dosage determines its effects. In these studies, males are put on testosterone blockers, so that their body's own production is brought to zero. Then the researchers add testosterone back in, at different amounts, while the participants follow a strength-training regimen. There's variation in the response to these manipulations, but the trend is clear: if you're training, more testosterone means more muscle and greater strength.

The effects of testosterone have been clear to athletes and their trainers for decades. Steroids, the scourge of competitive sports, are molecules built to mimic the chemical structure and biological effects of testosterone.* These drugs don't grow muscle by themselves—like real testosterone, they only work if you exercise. Nonetheless, the effects are obvious. More artificial testosterone in the form of steroids means bigger muscles and more strength. Of course, it can also shrink the testicles, grow breasts on males, develop facial hair and deep voices in females, and cause serious problems with the heart, kidneys, immune system, fertility, and elsewhere. Like most cheating, even if you can get away with it, it's not a great idea.

Other genes affect athletic performance as well. One that has

---

* Testosterone is classified as a "steroid" hormone based on its chemical structure. Lots of hormones and chemicals could, technically, be called "steroids." Here I'm referring to anabolic-androgenic steroids, the testosterone mimics commonly used to cheat.

received a lot of attention is *ACTN3*, a gene that produces a protein, called alpha-actinin-3, used to build actin filaments. *ACTN3* is particularly important in fast-twitch, type II muscle fibers. One variant, called the *R* allele, appears to promote muscle response and recovery after exercise. The other, called the *X* allele, doesn't appear to have any special effect. Neither appears to provide a major advantage in normal daily life; both alleles seem to be neutral. Worldwide, *R* and *X* alleles are equally common, and there's no evidence that selection has favored one over the other. However, in the strange, modern world of elite sports, at the highest levels of competition, a small but important effect emerges. When we look at type II specialists, like sprinters and powerlifters, the *R* allele is much more prevalent than we'd expect by chance. Nearly every power athlete at the Olympic level carries one or two copies of it. It seems that over years of training and competition, those without the *R* allele are at a disadvantage and tend to be winnowed out.

Another gene associated with athletic performance is *ACE*. Unlike *ACTN3*, the *ACE* gene is active in a variety of tissues throughout the body, and is involved in systems like blood pressure regulation that aren't obviously connected to sports. The evolutionary pressures acting on the *ACE* gene most likely revolve around these essential functions, not athletic performance. However, as with *ACTN3*, when we look at elite athletes a curious pattern emerges. The *I* allele of the *ACE* gene is more common among sprinters than we'd expect from chance, and the *D* allele is more common in endurance athletes. The precise cellular mechanisms underlying this apparent performance effect have yet to be resolved.

ACE and *ACTN3* are hardly alone. Twin studies in the 1990s indicated that the proportion of fast- and slow-twitch muscle you start out with is heritable, which in turn could affect whether you're inherently built more for power or endurance. These studies also showed heritability in the maximum rate at which your lungs can pull in oxygen and distribute it to your muscles, called VO_2max, a key trait for endurance. We've seen in earlier chapters that twin studies tend to overestimate genetic influence, but there's little doubt that there is a genetic component to physical ability. GWAS and other modern techniques have steadily increased the number of genes implicated in strength and endurance. Some alleles even affect your *response* to exercise, whether your body gets fit quickly or is difficult to budge. Like height, weight, IQ, and other complex traits, there are no doubt hundreds, probably thousands, of genes that influence your athletic ability, each with a nearly imperceptible effect.

It's also likely that genes influence your interest in athletics and physical activity. GWAS studies have identified alleles associated with how much people exercise—or at least, how much they say they exercise, since these studies generally rely on surveys to assess exercise frequency. Like the GWAS studies we discussed with IQ, there are a number of caveats and limitations. The predictive power of genetics is very small, for a start. Plenty of people with lots of "exercise" alleles are lazy, and many without those alleles are active. And as with IQ, we don't really know what organs or functions are actually involved. Alleles that promote exercise could be involved in neurotransmitter signaling that make exercise feel good, or the muscles' ability to recover and feel fresh, or personality traits associated with competitiveness, or a range of other things. Perhaps alleles that accelerate your growth, making you the tallest kid in sixth grade and the star of the basketball team, track you toward a life of enjoying sports and seeking out physical activity.

Whatever the case may be, the confluence of genetics and environment, nature and nurture, is right there staring us in the face every time we look in the mirror, head to the gym, or contemplate a run. By far the biggest factor shaping your muscles and your fitness is the time you invest in exercise. You can probably trace some of your interest in exercise to your DNA, but the environment you grew up in and surround yourself with today has a much greater effect. The way your body responds to exercise is mediated by your biology, the hormones you produce and, to some extent, the alleles of genes like *ACTN3* and *ACE* you carry. But unless you're an elite athlete, spending hours each day training, you can bet on two things: your abilities owe far more to your lifestyle than your genetics and you could probably benefit from more exercise.

FASTER, HIGHER, STRONGER: THE PHYSIOLOGY OF GREATNESS

Fifteen miles per hour (24 kph) is not a particularly inspiring speed. Unless you're looking for parking at the mall, it's frustratingly slow in a car. On a bicycle, it's just enough to feel the wind in your hair. No one bothered to record the first time a human ran 15 miles per hour, probably because it occurred in the Paleolithic period, before the invention of miles, hours, and the number *15*.

But in the years following World War II, the world of track and field was absolutely electric with the possibility of someone running 15 miles per hour for a full four minutes. Manage that, and you'll run a mile in four minutes flat. Just a smidge faster, and you break the mystical four-minute mile. During the early days of the space race, a sub-four-minute mile was the moon shot of distance running.

Several athletes had gotten tantalizingly close. Prior to the war, in 1937, British runner Sydney Wooderson had set a world record of 4:02.6, an average speed of 14.83 miles per hour. During the war years, while men from Britain and other parts of Europe were sent to fight, Swedish runners Arne Andersson and Gunder Hägg traded the record back and forth as Sweden remained neutral. By 1945, Hägg held the record at 4:01.4 (14.91 mph).

Roger Bannister was too young to fight in the war, just thirteen years old when his family sheltered in their basement in the city of Bath as the bombs from a German air raid rattled the house above them. Bannister had success running cross-country in high school, and he continued to run and compete in the 800 meter, 1500 meter, and mile as he pursued a degree in medicine at Cambridge after the war. Despite his obligations at school and notoriously light training regimen, he was among the elite at these distances through the late 1940s and early '50s. After a disappointing fourth-place finish in the 1500 meter at the 1952 Olympics, Bannister mulled retirement, but decided instead to focus his energy on the four-minute mile.

Two years later, in May of 1954, in a meet at the Iffley Road track in Oxford, Bannister broke through. In front of a crowd of three thousand, with the help of two pacers for the initial laps, he ran 3:59.4 (15.04 mph). The four-minute mile had finally fallen.

The classic explanation for elite performance like Bannister's is *talent.* Some people just seem born to excel in a particular activity. Francis Galton, the grandfather of the eugenics movement, spent much of his career in the late 1800s tracking successful people and their offspring. A polymath, Galton even developed the mathematics of correlation and regression, cornerstones of modern statistics, to perform these analyses. He was convinced, as any eugenicist must be, that excellence is fundamentally genetic, that some people are simply better than others. Galton would have no doubt agreed that *some* practice was necessary to develop one's potential. But in his view—the prevailing view through much of the 1900s, and still popular today—the main difference between the elite and the everyday was genetic, innate, and immutable.

A gene-centered view of excellence makes a number of predictions about the way that performance should vary among people and over time. First, for someone like Bannister who dedicates himself to a single pursuit like running, their performance ought to reach a plateau—their personal genetic potential—after a few years of training. Second, and relatedly, it shouldn't matter too much when a person gets started in a sport or other activity. Since it only takes a modest amount of practice to uncover greatness, those destined to excel should reach the upper echelons in their twenties regardless of whether they started in kindergarten or high school. And finally, when we track elite performance over generations, we should see similar levels over many decades. After all, genes evolve very slowly, and the same alleles that were mixing around in our great-grandparents' generation are still here today.

It turns out that every one of these predictions is wrong. When we look at the performance of individuals like Bannister, Andersson, and Hägg, we see continuous improvement year over year, at least until age catches up with them. In less physically demanding tasks, like playing darts* or the violin, elite performers improve throughout their long careers. Because your abilities depend on all the practice you've accumulated, it pays to start young. From Itzhak Perlman to Tiger Woods, elite performers begin in childhood, often as toddlers. And when we look over decades, we see performance continuously progressing as each generation finds better ways to train. Bannister breaking the four-minute mile, amazing as it was, was just another step in the long march of improvement in the men's mile from the early 1920s through today.

K. Anders Ericsson, a psychologist with expertise in expertise,

* Darts fans will know Phil "The Power" Taylor, who dominated world-championship darts for three decades, with some of his best performances in his late fifties.

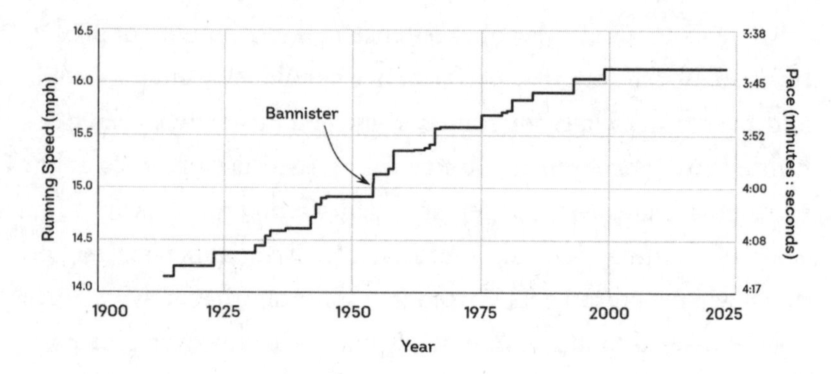

Figure 6.2 World records for men's one-mile run. Bannister's record (indicated with an arrow) was short-lived.

synthesized the science of excellence in a foundational paper in 1993. He and his colleagues examined the evidence from dozens of studies investigating performance in areas as diverse as swimming, playing violin, distance running, and typing Morse code. Everywhere they looked, modern standards far outpaced those in the past.* The world of sports is full of examples. Darts, archery, swimming, bowling . . . the standards everywhere increase generation after generation. Olympic champions in the 1896 marathon would barely qualify for entry into the Boston Marathon today. Even in artistic pursuits that don't require strength but put a premium on muscle coordination and speed, the best continually get better. Elite violinists in the 1800s balked at playing Tchaikovsky's concertos, considering them too difficult. Now they're part of the standard repertoire.

Another common theme was the importance of training. Consistently, across domains, it took ten years of serious, deliberate practice with good teachers to reach the elite level. At twenty

* There are clear parallels here to the Flynn effect of rising IQs over the past few decades. Like race performance, test performance reflects environment and practice, not genetics. Apparently, people in the early 1900s were slower in every dimension.

hours of practice per week, that's roughly ten thousand hours. The work of Ericsson and his colleagues gave rise to the ten-thousand-hour rule, made famous by Malcolm Gladwell in his book *Outliers*. Greatness, they suggest, is available to anyone who puts in ten thousand hours.

Gladwell's book, like Ericsson and his colleague's paper, argues that talent is a myth. Prodigies are simply kids who practice a lot, either because they love it or because they're pushed by overzealous parents. The experience they accumulate early in life puts them ahead of their peers, steering them into programs with more and better training, further magnifying their childhood advantages. By the time they're young adults, what looks like genius on the soccer field or concert stage is really just the tip of the iceberg, shining in the sun, floating atop an unseen juggernaut of practice. The difference between the Bannisters and the also-rans isn't genetics or talent, it's ten thousand hours of hard work.

The ten-thousand-hour rule would seem to explain why my friend Joe Beimel grew up to be a major league baseball player and I did not. Joe and I are the same age, from the same no-stoplight town in rural Pennsylvania, went to the same public school, and played baseball in the same Little League. But when I got bored with baseball in my preteen years, Joe doubled down. It helped that his dad seemed just as excited about baseball as he was. By the time we graduated high school Joe had surely logged more than his ten thousand hours. He went on to a thirteen-year career as a closer in the major leagues, pitching for the likes of the Dodgers and Pirates. I went to grad school.

Joe would probably be the first to tell you that you don't reach the upper echelon of any sport without putting in thousands of hours, just like Gladwell and Ericsson argue. But if you've been lucky enough to see an elite performer develop up close, it's hard to dismiss the idea that something about them is different. I would argue that the ten-thousand-hour rule isn't wrong, but it's incomplete. There's a secret ingredient in the recipe for success, just a smidge, that seems to be a mixture of genetics and chance.

First, there are the basic physical attributes. Different sports favor different builds. You don't see many five-foot-tall basketball players or six-foot-tall gymnasts at the elite level. The mechanics of pitching favor players who are tall, and Joe is six foot three (191 cm), about four inches taller than the average American man. As we discussed in chapter 2, your genetics play a large role in determining your height. The alleles Joe carries for height undoubtedly improved his chances of becoming a professional baseball player.

Modern genetics is also revealing alleles that seem to give elite athletes a mental or physical edge. As we discussed above, alleles of the *ACTN3* and *ACE* genes appear to affect muscle function and response to exercise. Plenty of other genes tip the scales as well, each a tiny bit, leading to the heritability we see in traits like muscle type and VO_2max. And then there are the hundreds, probably thousands, of genes with tiny, indirect effects on other systems. For example, genetics influence developmental timing, and kids who grow and mature a bit faster might well find themselves the biggest or most coordinated player on their childhood teams, tapped by coaches as "natural athletes" and shepherded into a life in sports. Muscle coordination, which is primarily an issue of brain wiring, also seems to be heritable, with GWAS analyses showing small, but measurable, genetic influence on the ability to keep rhythm.

We also know that genetics influences personality, including your interest in physical activity. The effects are small (environment matters more), but they likely help to explain why some kids happily plow through ten thousand hours of practice, spending all of their free time swimming or skating or kicking a soccer ball. Having watched Joe develop as an athlete through grade school and high school, I'd be willing to wager that he's got more than his fair share of alleles that promote perseverance and competitiveness.

Joe has also been lucky. He's left-handed, for a start, and lefties are overrepresented among pitchers in professional baseball. About 25 percent of major league pitchers are left-handed, compared to only 10 percent in the general population. Handedness results from asymmetries in the way the brain develops, leading to better coordination from the motor control center on one side.* Its heritability is pretty low, typically less than 0.10 for GWAS studies, meaning that the quirks in brain development that led to Joe's left-handedness were mostly due to chance.

Chance puts its fingers on the scale in other ways as well. There's the element of luck in competition, the gut-wrenching twists of fate familiar to any fan, that can derail a promising career or give an overlooked benchwarmer an opportunity to shine. Often the impact of chance is more mundane. Your ten thousand hours of practice won't mean as much without good instruction, for example, so you had better hope you grow up near amazing coaches. Even your birthday can color your odds. Youth development leagues typically group kids by birth year, meaning that players born in January are generally older (and bigger, and more mature) than their teammates who were born in December. A twelve-month difference can mean a lot when you're young, providing more time for practice and growth. Those advantages compound over time and, as a result, the best athletes tend to have birthdays early in the year. We see this "relative age effect" in a wide range of sports, from ice hockey to soccer to track and field.

---

* You can retrain yourself to use your nondominant hand, as children in my parents' and grandparents' generations understood. We'll dig into handedness more in the next chapter.

The effects are strongest among young athletes, but in many disciplines they persist into adulthood.

UNDERSTANDING HOW THE BEIMELS AND BANNISTERS OF THE WORLD become elite athletes helps to explain why some countries and populations come to dominate particular sports. Athletic success has the same recipe wherever you go, but the ingredients are locally sourced and their impact depends on the scale of analysis. Genes influence an individual's abilities but don't offer a great explanation for the dominance of particular groups, because individuals are unique in the particular set of alleles they carry, yet groups around the world share over 90 percent of the same alleles. As far as we know, none of the alleles linked to sports performance are exclusive to any particular population. Chance impacts particular athletes and, through accidents of history, can also shape an athlete's environment. Athletes in India and Pakistan, for instance, excel at cricket in part because English colonists introduced the sport.

The patterns we see in international sports, the reason that Asian countries dominate in ping-pong, for example, or northern Europeans in the biathlon, come down to differences in environment. To contend for a medal, an athlete needs to grow up in an environment that gets them practicing from an early age. Kids gravitate toward whatever sports are available, especially those that the adults in their lives hold up as exciting and important. Norwegian kids find Nordic skiing, Japanese kids find judo, Jamaican kids find track. If you have thousands of children exposed to the sport from a young age, with a strong tradition of good coaching, the odds of fielding a national team of five or ten Olympic-level athletes every four years are pretty good.

In big, rich countries like the U.S., children have a broad range of sports to choose from, but opportunities and encouragement are uneven. It's hard to learn to ski if you don't live near snowy mountains, and the equipment, lessons, and lift tickets are pricey. If you grow up in a poor neighborhood without easy access to a swimming pool, or you don't have a parent who's free to spend their mornings driving you to swim practice, you're unlikely to get your ten thousand hours of the butterfly or backstroke. In addition to the obvious socioeconomic disparities in access, kids in the U.S. grow up immersed in a culture where, for example, Black athletes like LeBron James and A'ja Wilson dominate in basketball, while White athletes like Katie Ledecky and Michael Phelps dominate in swimming. It shouldn't surprise us that, year after year, our national swim team is mostly White, while our national basketball team is mostly Black.

That's not to say that genes don't play a role, or that the genetic playing field is even. It's very clearly not: some people carry alleles that give them an advantage. But the alleles implicated in elite performance are found in every population, every race, every country. These alleles might be more common in some populations than others, but with a large enough population to draw from there will always be plenty of eager kids who carry them. By the time that the millions of aspiring young athletes have been through the filter of youth competition, the handful who are left and find themselves among the elite are all carrying a set of alleles that help them excel.

Our tendency to notice patterns, and the undercurrent of genetic determinism in sports, leads us to overgeneralize and misinterpret the lopsided results we see in competitions like the Olympics. Some East African populations, like the Daasanach whom we met in chapter 3, carry alleles that promote a tall, thin build—body proportions adapted to hot climates that also happen to be well suited for endurance running. We shouldn't be surprised when the best athletes from these populations excel at the marathon, particularly when distance running has become a source of national pride. But that doesn't mean they have some unique genetic advantage that confers excellence in distance running. There are plenty of tall, thin, successful marathoners in other populations as well. Paula Radcliffe, a White woman from the United Kingdom with a BMI of 18, held the women's world record in the marathon for fifteen years, from 2002 to 2017.

Similarly, every sprinter in the Olympics is likely to carry two copies of the *ACTN3 R* allele that promotes anaerobic power, regardless of whether they're Black or White. Swimming and track provide a useful comparison here. The 50-meter freestyle swim and 200-meter sprint are both anaerobic, power-based events that last about twenty seconds. White athletes have historically dominated in the pool, Black athletes on the track. The apparent racial difference has nothing to do with genetics or physiology. Both events ask the same from your muscles. The alleles that help you sprint in the pool are equally as important on land. The racial divide in swimming and track comes down to culture, environment, and opportunity. No race has an innate, genetic advantage over another.

As diversity in sports grows and opportunities expand, I'm hopeful that racist ideas about sports performance will continue to die away. We saw glimmers of this early in the modern Olympic era, with Jim Thorpe, a Native American, winning gold medals in the pentathlon and decathlon in 1912, and Jesse Owens, a Black man from the U.S., winning four gold medals in sprinting and the long jump, embarrassing Hitler and the other eugenicists in the 1936 Games in Berlin. In the decades after World War II, even as professional sports in the U.S. slowly integrated Black athletes, managers and coaches resisted the idea that Black football players could be great quarterbacks. Today, athletes like Patrick Mahomes and Jalen Hurts, the quarterbacks who faced off in the Super Bowl in 2023, have proven otherwise. Tennis was a traditionally White sport until athletes like Arthur Ashe and the Williams sisters hit the court. Basketball in the U.S. has been dominated by Black athletes for decades, but two of the best in the NBA today, Luka Dončić and Nikola Jokić, are White guys from Eastern Europe, and Caitlin Clark, a White woman from Iowa, recently broke the all-time collegiate scoring record. The gold medalist in the 100 meters at the 2020 Olympics was Marcell Jacobs, an Italian runner with a White mom and a Black dad.

As the podiums and record books become more diverse, hopefully people's attitudes toward excellence in sports will continue to evolve. In the meantime, we should be conscious of the stories we tell about the athletes we revere. There's no harm in recognizing the dominance of Jamaican sprinters or Ethiopian marathoners, and it's not necessarily racist to expect the Chinese ping-pong team to bring home a gold medal. But we do those athletes a huge disservice (and we wade into racist waters) when we chalk their performance up to race or genetics. Like all athletes, they win because they do the work. Their race has nothing to do with it. To suggest otherwise is to dismiss the ten thousand hours and more that these athletes have devoted to their sport.

Stories about race aren't the only ones we tell about our bodies. We've touched on sex differences in sports in this chapter, but there is much more to discuss about male and female physiology.

7
TURNING ENERGY INTO OFFSPRING

Hadza arrow poison is strong stuff. As teenagers, young men learn the secret recipe, passed down from fathers and uncles through the generations. Chopping limbs from the smooth-barked panjube tree into small chunks, boiling them over a fire to extract and concentrate the caustic sap, forming the black-brown putty into a smooth slug around the base of the arrowhead, taking care not to poison themselves. An arrow fired from a handmade Hadza bow leaves the giraffe-tendon string at well over 100 miles per hour (more than 50 meters per second), with enough force to rip right through the rib cage of a warthog or plunge deep into a kudu's flank. It's not the physical damage that dispatches the quarry, however, not with large game. It's the poison. That one-ounce slug, absorbed into the blood, can kill a nine-hundred-pound (400 kg) zebra within hours.

Imagine my surprise, then, as I was walking out of camp one morning with Mwasad, a Hadza man and prolific hunter, when a woman came running up to catch him before he left. After a short, private exchange, he handed her his bow and quiver of arrows and she walked off. *Change of plans.* Mwasad turned back to camp, his body language making it clear that we wouldn't be hunting today. There would be no point. His arrow poison wouldn't work.

What could cause such a fundamental chemical change? For Mwasad, the answer was obvious: a woman. His wife had just started menstruating. It was common knowledge that a wife's menstrual blood will ruin her husband's arrow poison if he's foolish enough to go hunting during her period. She didn't need to handle the arrows or even be near them for this to happen; his connection to her as her husband was sufficient to cause the damage. Even *touching* his bow and arrows was a risk, as it could render the poison inert. Their friend had saved them that morning, relaying the news to Mwasad and collecting his gear, hopefully before anything bad had happened.

There are two lessons here. The first, of course, is that you can blame your spouse for *anything*. The second, equally universal, is that we all live our lives confined within a Story of How the World Works. Like a fish in water, this story envelopes us so completely that we don't even know it's there. It's a tapestry of interwoven theories, covering every aspect of our lives: everyday physics, what counts as food, who's family, friend, and foe, how a good person is supposed to act, why the leaves turn colors in the fall, where the sun goes at night, where babies come from and how to raise them, what happens when we die, the nature of the supernatural.

The Story of How the World Works is the foundation of any culture, the hard-won wisdom of the community. We learn it so early, absorb it so completely, that it seems obvious and infallible. Only when we have the chance to observe it in another community do we notice how much is arbitrary and strange. These stories have developed independently in different populations over millennia, with people doing their best to craft a coherent framework of rules from a world that's often confusing and unpredictable. Some elements are universal. No communities try to make fire from wet wood, most are wary of snakes. Beyond the basics, things get creative. With no understanding of bacteria or viruses, traditional stories around disease—who gets sick and why, how to cure the ill—often invoke magic, the stars, or the gods. It's remarkable that the Hadza figured out how to make arrow poison without the foundations of modern chemistry. Their theory of how the poison works and the imagined connection to the menstrual cycle speaks to the human need to see patterns, find explanations, and create rules for things we don't really understand. It may be arbitrary and incorrect, but for the Hadza it's as obvious and real as water flowing downhill.

Nowhere in our daily lives is the Story of How the World Works more prominent than in the cultural norms around sex and gender. Across time and throughout the world, cultures everywhere delineate male and female, with gender-based rules, expectations, and obligations that govern every aspect of one's life. The Hadza certainly have theirs, and they go well beyond arrow poison. Men are allowed to have multiple wives, but women can't have multiple husbands. Hadza men and women spend their days apart, in men's or women's groups. Homosexuality isn't discussed or acknowledged. Hunting and honey collecting is men's work, gathering tubers and berries is for women. When large game is taken, only men can eat the heart, kidneys, and testicles.

The variety of customs and beliefs about sex differences, gender, and sexual behavior will renew your faith in human imagination. Some groups like the Hadza put men and women on equal footing, while others like the Daasanach have a strongly enforced patriarchy. A few put women in charge. Gender in many societies is limited to male and female, but others acknowledge and even revere transgender or third gender individuals. People have sex in every mechanically feasible way with any variety of partners, but cultural rules about who and what are permitted on the sexual menu are as diverse as local cuisine.

In the last century, the U.S. and many other societies around the world have witnessed a massive reconsideration of male and female differences, sexual behavior, and gender identity, and whether the old rules and obligations are consistent with a modern sense of fairness and inclusion. Traditionalists and the progressives have both taken up the mantle of science, the modern authority on How the World Works. Traditionalists point to differences in size, chromosomes, hormones, and genitalia, arguing that age-old ideas about sex and gender reflect an immutable, natural order, rooted in biology. Progressives point to the overlap in male and female physiologies and abilities, and the scientific evidence that environments shape our differences, arguing that the distinction between male and female is cultural, arbitrary, and flexible.

It's an existential clash between competing Stories of How the World Works with real-world impact on people's rights and daily lives. It's also deeply personal. As long as we're not hurting anyone, the way we live our lives, feel about our bodies, and express our sexuality doesn't seem like it should be up for public discussion. And yet, a heated public discussion of sex and gender is precisely what we've got. Human biology figures prominently on both sides of the debate. To weigh the arguments, or just to inform our own worldview, we need to understand how male and female bodies develop and function.

Some house rules as we wade into a contentious area. People's preferred vocabulary around sex and gender varies and changes, and so I know the discussion that follows might not use the terms you'd choose. My goal is to give a clear account of human reproductive biology without causing offense or relying on unfamiliar language. I will use "male" and "female" to refer to chromosomes, anatomy, and physiology, using the familiar distinctions of biological *sex* (e.g., males typically have XY chromosomes, a penis, and higher testosterone levels). "Biological mom" and "biological dad" refer to the individuals whose egg and sperm respectively produce an embryo. "Biological mom" or "mother" gestates the pregnancy. I'll reserve "boys" and "girls" or "men" and "women" for discussions of *gender*, which refers to the way people understand themselves and are seen by others.

The distinction between sex and gender is relatively recent, largely a product of feminist scholarship in the 1960s and '70s. Some argue the distinction was an unhelpful invention born of political correctness, others that it's already out-of-date and too conservative. Many who study human biology, including me, find it helpful for talking about our bodies, behaviors, and experiences. Another favorite front in today's culture wars is the battle to define the word "woman." We'll take that on when we get there.

CONSTRUCTING SEX

Imagine yourself at sea, exploring uncharted islands in the vast Pacific. Land appears on the horizon, two green peaks jutting up from blue waters, side by side. Excited by the new discovery, you captain the ship around the perimeter. From the east or west, both peaks are clearly visible, but from the north or south the view of one overlaps the other, making it hard to tell them apart. You send a landing party for a detailed inspection. Upon their return, the cartographer tells you the peaks are clearly distinct, but not truly separate. There's a narrow isthmus connecting them. The geologist isn't surprised. She points out that both peaks sit atop the same enormous seamount, unseen beneath the water. The biologist notes that the flora and fauna on either peak is similar, though some species are more common on one than the other.

Later, in your cabin, you sit down to record the discovery in your logbook. But what story do you tell? How do you frame it? Is the formation one island or two peaks? Do you focus on the differences, the clear separation that you can see when you sail in from the west, the subtle but interesting differences in ecology? Or do you concentrate on the unity, the complete overlap you find approaching from the north, the shared geology and foundation, the clear connection?

The most honest approach is to faithfully describe *all* of it, the differences and the similarities, as accurately as you can. Both are

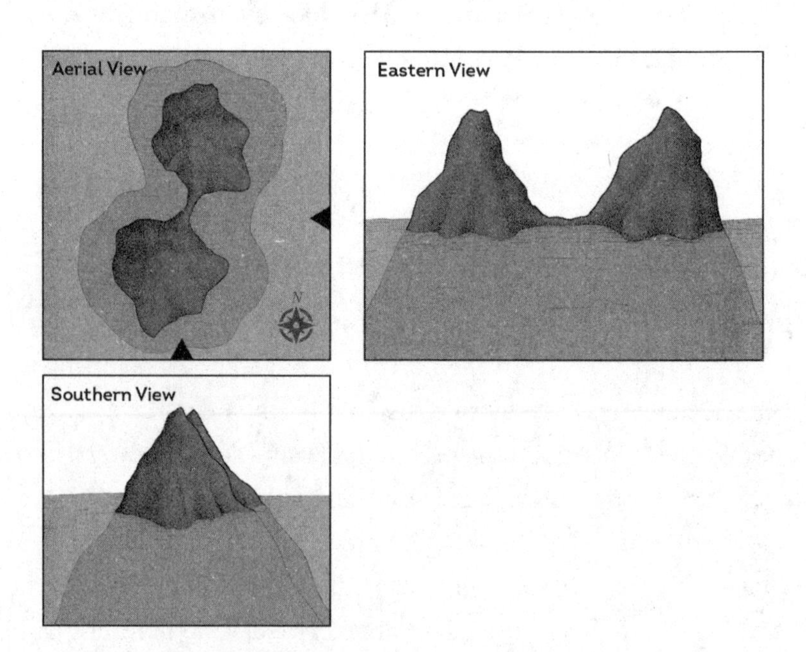

Figure 7.1 Peaks rising above the Pacific Ocean, connected by an isthmus and resting atop a shared seamount. From above, it's clear they are both distinct and connected. The degree of similarity versus separation depends on one's view.

true, even if they seem contradictory. So it is with the biology of male and female bodies.

Our bodies start out sexless. Up until the sixth week of embryo development, the tissues that will form the reproductive system are wholly ambiguous. Undifferentiated gonads, neither male nor female, sit alongside the early kidneys and some generalized ductwork. Barring any interference, by the seventh week the gonads begin the transition to ovaries, while the ductwork forms the fallopian tubes, uterus, and vagina. The external anatomy develops into a vulva, including the labia and clitoris. Everyone with a typical human genome—male and female alike—carries the genes required to produce the female reproductive anatomy; it's the default setting for mammals. (That, incidentally, is why everyone has nipples—they are part of the standard mammalian body plan.)

Under the influence of a functioning *SRY* gene, located on the Y chromosome, the undifferentiated gonads chart a different path, developing into the male anatomy. Humans typically have forty-six chromosomes, which are really twenty-three pairs: one set from biological mom and the other from biological dad. One of those pairs is known as the sex chromosomes. In humans, as in other mammals, there are two kinds, X and Y. The *SRY* gene becomes active around the sixth week of development, producing the SRY protein, which kicks off a cascade of other gene activity and changes to the developing reproductive anatomy. The gonads develop into testes.

Hormones produced by the testes, including testosterone, cause the ductwork to develop into the tubes (vas deferens) that carry sperm from the testes to the urethra (the exit tube that runs the length of the penis) as well as secondary structures like the prostate. The external anatomy develops into a penis and scrotum. The tissue that would have otherwise formed the clitoris becomes the penis, while the tissue that would have become the labia forms the scrotum.

Given the role of the Y chromosome in sexual development, biologists use the term "chromosomal sex" to describe your particular assortment. "Chromosomal females" have two X chromosomes, one they inherited from biological mom and the other from biological dad. "Chromosomal males" have an X chromosome they inherited from biological mom and a Y chromosome they inherited from biological dad. (Biological mom can't pass on a Y chromosome because she doesn't have one.)

Like everything else with sex, these clear distinctions can get complicated. Missing or extra chromosomes are rare, but occasionally occur in the formation of a sperm or egg, resulting in an embryo with something other than the usual forty-six. Down syndrome, for example, occurs when an individual has three chromosome 21s. Similar situations can occur with the sex chromosomes. Individuals can be XXY (known as Klinefelter's syndrome), XYY (Jacobs syndrome), or XXX (triple X syndrome). You can even be X (Turner's syndrome), with a single X chromosome that lacks a partner.* People who are XYY or XXX often develop normally, and might not even know they're affected. XXY and X are typically more disruptive and can cause a range of problems, including reproductive anomalies and infertility.

The complications don't end with unusual chromosome assortments. Even among XX or XY individuals, some people end up with intersex anatomy that doesn't align with their chromosomal sex or is difficult to distinguish as either the typical male or female form. The ambiguity arises from the way our reproductive systems are built. Male and female anatomy develop from the same

* Embryos don't survive and develop without at least one X chromosome, as it carries too many essential genes—most of which are unrelated to sexual development. As a result, no one is born with only a Y chromosome.

starting materials, and develop into the female form unless they receive a large dose of testosterone and related hormones. Those hormones are called "androgens," Greek for "male making." The cells they target need to have functioning receptors, docks for the androgen molecules to bind to, for androgens to have an effect. When the production of androgens or their receptors goes awry, a range of outcomes can result.

In "congenital adrenal hyperplasia," chromosomal females (XX chromosomes) produce unusually high levels of androgens from their adrenal glands—the main source of androgens in females. Sometimes, the high dose of androgens can cause the genitalia to be ambiguous or to appear male, the clitoris looking like a penis, and some of these individuals are raised as boys. In rare circumstances, the *SRY* gene is copied onto the X chromosome during the production of a sperm cell. Despite its atypical location, the *SRY* gene exerts its usual developmental effects, resulting in a person with XX chromosomes and male genitalia. Conversely, people with XY chromosomes can develop female anatomy if their *SRY* gene isn't functional. In "androgen insensitivity syndrome," chromosomal males develop testicles (their *SRY* gene and hormone function normally), but the receptors for testosterone and other androgens aren't fully responsive. Depending on the severity of the receptor problem, their genitalia may develop into the female form, with a vulva and vagina. In a condition called 5-ARD, a chromosomal male produces testosterone and has functioning receptors, but an enzyme needed to convert testosterone to another androgen, called DHT, is missing. DHT is critical for the development of the penis and scrotum. Without it, individuals with 5-ARD can develop female genitalia.

The list goes on. There are a number of developmental detours that can lead to intersex anatomy, and together they affect approximately one in five thousand newborns each year. These millions of people are the isthmus of our imagined Pacific island, the thin but very real connection between the typical male and female anatomies. There's a range of other perturbations and problems that can occur in building a reproductive system, just as there are with any of our body systems. When we include these conditions, the rate of "differences of sexual development" increases to around 1 in 250.

For anyone keeping track, we're now up to two definitions of sex: chromosomal sex and anatomical sex. Most of the time they are unambiguous and aligned. Occasionally they are not.

ADULT SWIM

Like our imaginary mountains rising from the sea atop a common platform, we can see how the developing male and female anatomy are really just variations on the larger, shared foundation of reproductive anatomy. Testicles and ovaries, labias and scrotums, they all develop from the same parts. It makes sense, then, that they share the same hormonal control. In both males and females, the hypothalamus in the brain produces the hormone GnRH, which stimulates the pituitary gland resting just beneath the brain to produce two hormones, LH and FSH. Those hormones stimulate ovaries to produce estrogen and progesterone, and they also stimulate testes to make testosterone.

There are three hormone tsunamis that wash through our bodies over the course of development. The first happens when you're still in the womb, as testosterone levels in males and estrogen levels in females reach the highest concentrations the body will ever experience. These are the hormone surges responsible for building your reproductive anatomy. In females, these surges also lead to the production of eggs. Females are born with all the eggs they will ever have already packed into the ovaries, waiting in a sort of suspended animation until adulthood.

The second wave of hormones is considerably smaller and occurs in infancy, peaking when you're one to three months old. The impact of this second surge isn't entirely clear yet, but it appears to influence the development of the genitals, overall body composition, and possibly the brain and behavior.

The third wave is the one that every person who survived middle school is intimately familiar with: puberty. Hormone concentrations in the blood climb to one hundred times the levels in late childhood. The effects, as you're probably aware, are widespread. There are the physical changes, of course, from pubic hair and genital development to the growth spurt in height and weight. Testosterone causes the vocal folds in males to thicken, lowering their voice, while also promoting the growth of bones and muscles. The testes mature, drop into the scrotum, and begin producing sperm. Rising estrogen levels in females lead to the development of breasts and the beginning of ovarian cycling, marked by menarche, the first menstrual period. Behaviors change as well. Brains in both males and females are affected by the flood of hormones, with an awakening of a person's sex drive, an increased interest in social connection (particularly with other adolescents), and what psychologists call "sensation seeking," pursuing high-intensity experiences.

Once through the gates of puberty, the reproductive system is in adult mode, able to do the things necessary to make more humans. A steady drumbeat of GnRH pulses from the hypothalamus, leading to FSH and LH production by the pituitary, which stimulate the testes or ovaries. In males, the testes produce testosterone, which has effects throughout the body. One of its effects is to quiet the production of GnRH, FSH, and LH. We've seen this sort of negative feedback system before, in chapter 2 with the production of growth hormone. That negative feedback—testosterone shutting down its own production—keeps testosterone levels in a

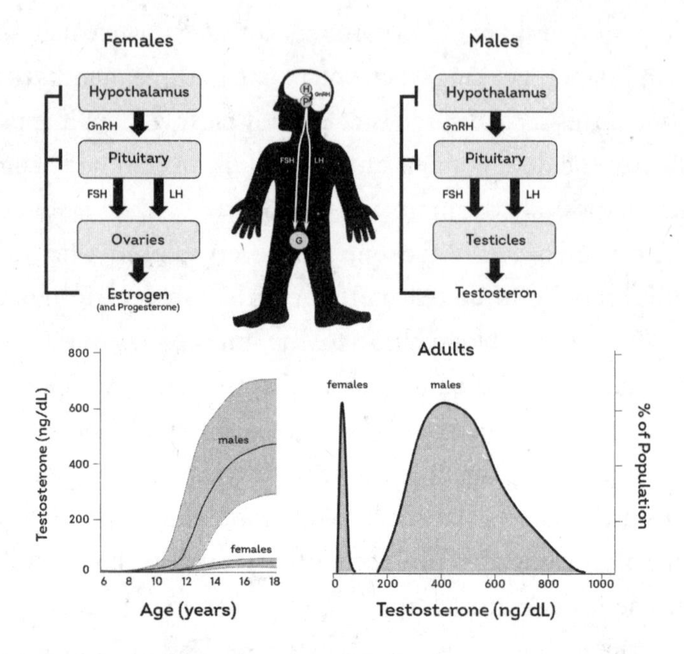

Figure 7.2 The hypothalamus (H) produces GnRH, which stimulates the pituitary gland (P) to make FSH and LH, which stimulate the gonads (G) (i.e., ovaries or testicles). Ovaries produce estrogen in the first two weeks of the ovarian cycle and then a mix of estrogen and progesterone during the two weeks after ovulation (they also produce some testosterone and other hormones). Testicles primarily produce testosterone in response to FSH and LH (they also produce some estrogen). In both sexes, the hormones produced by the gonads inhibit the hypothalamus and pituitary, an example of negative feedback that keeps hormone production in check. Hormone production and regulation matures at puberty. For children under ten, testosterone levels are low and males and females overlap. At puberty, males begin to produce far more testosterone. By adulthood, there is no overlap in testosterone levels. Estrogen production in females shows a similar increase at puberty.

narrow range, typically between 200 and 800 nanograms per deciliter of blood.* Testosterone isn't the only hormone made by the testes. They also produce DHT, a powerful androgen, as well as estrogen and a few others.

* This feedback is also why taking steroids, which are testosterone mimics, can shrink your testicles. All that fake testosterone shuts down GnRH, FSH, and LH production, the testes stop making their own testosterone, and they wither away.

The testes also produce sperm, tens of millions of them each day. Each has twenty-three chromosomes, the result of some intricate shuffling and sorting of chromosomes during meiosis, the specialized cell division that produces sperm in the testes and egg cells (or "ova") in the ovaries. To start things off, all forty-six chromosomes line up, each with their partner, in twenty-three pairs, inside a sperm- or egg-making cell. The chromosomes then trade chunks of DNA with their partners, a process called recombination. A section of the chromosome 15 from biological mom will swap places with the same section of the chromosome 15 from biological dad, for example, like athletes trading jerseys at the end of a game. This happens for hundreds of sections across all the chromosomes.

The resulting forty-six chromosomes will each be a mosaic of DNA from the person's biological mom and dad. This scrambling of DNA is, we think, one of the main reasons sex evolved in the first place. It creates genetic diversity in offspring, new and never-before-seen combinations of the DNA from biological mom and dad. Boosted genetic diversity, in turn, helps those offspring stay a step ahead of bacteria and other single-cell nasties that are continually evolving to pillage our body's resources and make us sick.

Having created forty-six newly shuffled chromosomes, the cell divides, pulling one set of twenty-three chromosomes into each of the resulting cells. (Occasionally, this sorting is uneven, resulting in extra or missing chromosomes.) Those cells then develop into sperm or eggs, each with twenty-three chromosomes. When a sperm cell combines with an egg cell the resulting embryo will have the full human complement of forty-six.

Delivering the sperm to an egg typically involves an activity called "sex." If this is the first time you're hearing about this, it's going to sound pretty wild. Aroused by a prospective mate, the spongy tissue in the shaft of the penis becomes engorged in blood, stiff and erect. If the mate is female, sexual intercourse might occur with the penis inserted into the vagina. Reaching a neurological climax called orgasm, the male ejaculates a couple milliliters of semen from the penis, a mix of fluids from the prostate and other supporting glands and millions of sperm from the testes. Those sperm, with their frantic, whiplike tails, will swim through the vagina and uterus, toward the fallopian tubes, pulled along by contractions in the female reproductive tract. If one or more happen upon a freshly ovulated egg, fertilization might occur (proceed to chapter 1). The whole thing is a rather unlikely Rube Goldberg setup, yet it has been enormously successful for over 200 million years of mammalian evolution and occurs over 150 million times in humans around the world each year.

If this is difficult to picture, there is an archive of images, video, and other descriptions occupying roughly half of the internet. But consider yourself warned. Learning about sex from the internet is a bit like learning about dinosaurs from the *Jurassic Park* movies. While the characters are based loosely on a real species, much of their behavior and anatomy is fictional, and the acting is uneven.

Hormone production on the female side is a bit more complex than in males. GnRH still stimulates the production of FSH and LH, and those hormones do the job of stimulating the ovaries. But the response of the ovaries changes over the course of the menstrual cycle, which is typically around four weeks long. The first half of the cycle, called the "follicular phase," begins when the female gets her period, which reflects the shedding of the blood-rich lining of the uterus. For the next two weeks, FHS and LH stimulate the ovaries to produce estrogen, and levels climb as follicles in the ovaries develop and grow. The uterus grows a new lining, in anticipation of a possible pregnancy.

Around day fourteen of the cycle, after a spike in FSH and LH, one of the follicles pops out an egg, a process called ovulation. This marks the beginning of the "luteal phase." For the next two weeks, FSH and LH will cause the follicle that ovulated to produce the hormone progesterone. Estrogen levels are maintained as well, a bit lower than they were late in the follicular phase. The reproductive system is prepared for a pregnancy. Around day twenty-eight of the cycle, if there's no sign of a pregnancy, progesterone and estrogen levels drop and the cycle starts again. If a pregnancy *does* occur, progesterone levels will remain high throughout gestation. The combination of progesterone and estrogen shuts down ovulation in order to prevent another pregnancy from starting when one is already in progress.

Infants can also prevent, or at least delay, another pregnancy from taking place, and not just by making sex nearly impossible for their exhausted parents. If a pregnancy is successful and the mother nurses her newborn, the hormonal control of milk production can shut down ovulation. A baby's suckling stimulates the hypothalamus to make two hormones, oxytocin and prolactin. Oxytocin causes the mammary glands in the breast to eject milk,* while prolactin stimulates them to make more milk for future feedings. Prolactin also disrupts GnRH production, which in turn suppresses FSH and LH production and the resumption of ovulation post-pregnancy. Both pregnancy and nursing require a lot of time and energy from the mother's body, and the system seems to have evolved to prevent a new pregnancy from starting too soon, before the mother has had a chance to recover from the last one.

It took years of effort throughout the mid-1900s to sort all of this out. The impact on reproductive behavior and family planning has been nothing short of revolutionary. A working understanding

---

* Oxytocin is also wrapped up in the emotional connection between mother and baby, and can be stimulated by the sound of a baby crying or other infant-related stimuli, causing untimely milk release.

of ovarian hormones enabled researchers to develop the first highly effective oral contraception. The Pill, as it's still widely known, is a combination of synthetic estrogen and progesterone that mimics hormone levels in the luteal phase and pregnancy. It prevents pregnancies by shutting down ovulation, just like those hormones do when produced by the ovaries. Every female contraceptive drug on the market today, whether it's a pill, an injection, or a subdermal implant, still uses this approach (although some have no estrogen). Mifepristone, a drug used in medication abortions, blocks the effects of progesterone, effectively ending a pregnancy.

Other breakthroughs have helped people get pregnant. The hormone regime administered during fertility treatments is essentially a supercharged version of the follicular phase, designed to produce multiple ovulations and several eggs for fertilization. The FSH spike prior to ovulation provides the signal that at-home ovulation kits use to indicate the timing of peak fertility. At-home pregnancy tests look for a hormone called hCG, which is produced by the placenta to signal to the female that a pregnancy is in place.

Contraceptives and pregnancies aren't the only waves that can rock the usual menstrual cycle. The energy and time investment of pregnancy is massive, particularly for humans with our long gestations and helpless newborns. The female reproductive system has evolved to be cautious. If food is scarce, or the daily workload too demanding, the menstrual cycle might shut down: better to wait for improved conditions than to start a costly pregnancy that you can't sustain. In our Paleolithic past, this no doubt helped mothers make it through hard times. Today, we see female athletes and others with demanding exercise workloads, or females who are starving from lack of food or disordered eating, experiencing menstrual irregularity or fertility problems. Even high levels of stress in our modern world can negatively impact fertility.

The biggest disruption to menstrual cycling is menopause,

when the ovarian carousel stops completely. Menopause usually occurs between ages forty-five and fifty-five. Estrogen and progesterone levels both drop, causing a broad range of symptoms from hot flashes to thinning bones. Our understanding of hormones has helped here, too, with many benefiting from hormone replacement therapy, which typically involves doses of estrogen.

Male hormonal control is boring, by comparison. High exercise workloads can reduce testosterone levels and libido, but completely shutting down sperm production is rare. There's also a decline in testosterone levels through adulthood and old age, which coincides with lower libido and some decline in sexual function. About 40 percent of males in their forties experience erectile dysfunction, and the prevalence increases about 10 percent per decade after that. The main problem seems to be vascular (not testosterone levels), involving the blood flow and vessels in the spongy tissue of the penis.* High blood pressure, which damages vessel walls, is a risk factor, another reason to exercise and keep your heart and blood vessels healthy. But aside from age effects, the male reproductive system is pretty stable throughout adulthood, producing sperm and otherwise doing its jobs into the golden years. Actor Al Pacino recently fathered a child at the age of eighty-three. There are documented cases of fathers in their nineties.

SEX IN THE STONE AGE

The effect of sex hormones in our bodies goes way beyond our reproductive anatomy. Just how far they reach, and how different they make males and females, has been fertile ground for evolutionary storytelling ever since Darwin.

---

* This would explain why sildenafil, the game-changing erectile dysfunction drug sold widely under the name Viagra, was initially developed as a blood pressure treatment.

The traditional "Man the Hunter" view, led these days by the Paleo bro contingent, paints a picture of macho cavemen hunting woolly mammoths and fighting over nurturing, plant-loving females. Those conditions, in their telling, led to the evolution of massive sex differences in bodies and behavior that are real and innate—differences that progressive, woke society foolishly or unfairly tries to extinguish. As with most pop-psych evolutionary storytelling, this view takes a kernel of biological reality and extrapolates it into fantasy, often to excuse (or encourage) men being jerks.

It's true that, in many species, males evolved a larger body size and aggressive behavior to compete with other males for mating opportunities. But compared to our primate kin, human sex differences are pretty small. In chimpanzees and bonobos, our closest relatives, males are about 20 percent larger; in gorillas and orangutans, males are *twice* the size of females. Ape males also sport big nasty canines. Human males are pretty unimpressive by comparison, just 10 percent larger than females on average, with plenty of overlap. Our canines do not intimidate. All of this speaks to *reduced* male-male competition in the human lineage. Few would argue that male-male competition has been extinguished in humans, and we don't need to look very hard to find it alive and well today, across societies. But it seems there's been selection for males to be nicer, less competitive, more cooperative.

There's also evidence for female competition in our species. Across vertebrate species, when we see traits that develop in only one sex, and that don't seem to be related directly to survival or reproduction, those traits are often involved in competition for mates. Larger body size for fighting is one example, but we also find traits that improve attractiveness. The bright red plumage of a male cardinal, for instance. Usually it's males that evolve these advertisements, and traits like facial hair and deep voices might be examples of such traits in human males (they could also be related competition with other males). Intriguingly, the human female pattern of fat deposition to build breasts, hips, and backsides during puberty is unique to humans—other primates don't do it—and seems to indicate an evolutionary history of female competition for mates (fans of television's *The Bachelor* know this already). Reduced male body size and the evidence for female competition tells us that the evolution of sex differences in humans is far more complicated than the simple caveman soap opera.

Still, there *are* real differences in the body size and strength between adult males and females. As mentioned in the last chapter, these differences mainly emerge during puberty and are clearly related to the effects of testosterone, which is about fifteen times higher in males from puberty onward. Ovaries produce testosterone as well, and there are conditions that can increase female levels, including congenital adrenal hyperplasia (mentioned above) and polycystic ovarian syndrome. However, neither these nor any other common conditions raise female testosterone levels into the male range. From the perspective of testosterone levels, the male and female peaks are clearly distinct, with no overlap.

Elegant studies isolating the effects of testosterone by carefully manipulating levels in people's blood have shown very clearly that testosterone builds muscle. Testosterone also increases circulating hemoglobin, the molecule we met in chapter 4 that carries oxygen in the blood, and it magnifies the effects of growth hormone, which we met in chapter 2. As a result, *average* size, strength, speed, and endurance are about 10 percent greater for males. *Maximum* performance, measured among elite athletes of both sexes, differs by around 10 percent as well, and is over 20 percent in some sports. We can see this in the distribution of running speeds from the Boston Marathon (see figure 7.3). The ranges for males and females overlap considerably, with plenty of females who are stronger than their male counterparts, but the average and extremes are clearly different.

Two wrinkles about testosterone's effects are important to know. For one, variation in testosterone level *within the normal healthy range for a person's sex* doesn't correspond to greater or lesser strength. That's because circulating testosterone is only one piece of the equation. Testosterone receptors matter as well, as do our genetics and training. The obsession over T levels in the Rogan-esque corners of Man Culture is generally misplaced—it's not the clear-cut indicator of virility many think it is.

The other thing to note is that the size, strength, and endurance advantages that testosterone imparts don't go away when testosterone levels fall. Its effects are what biologists call "organizational"—they change the body in ways that aren't easily undone. Studies using medication to lower testosterone levels to female-like levels have shown only small changes in muscle mass, strength, and performance. Even after twelve months of treatment the reduction is usually around 5 percent in nonathletes, and doesn't fully erase the male advantage. Few studies have examined these effects in athletes, but we would expect the decrease to be smaller if they maintain a training regimen that minimizes muscle loss. Just like you can't unbake a loaf of bread, you can't undo the effects of the testosterone wave that washes throughout the body during puberty.

Training also has an enormous effect on speed, strength, and endurance, as do the alleles we carry for traits like muscle fiber type and body size. That's why there is so much overlap in male and female size and performance. Differences in training can even account for some of the population-wide male advantage, since they're more likely to participate in sports. But the data show a clear sex difference, mediated by the testosterone, that can't be explained away by environmental effects. Think back to our imagined Pacific island: when we look at size, strength, and endurance, we're seeing the male and female peaks from a three-quarters view. One largely occludes the other, but we can clearly tell them apart.

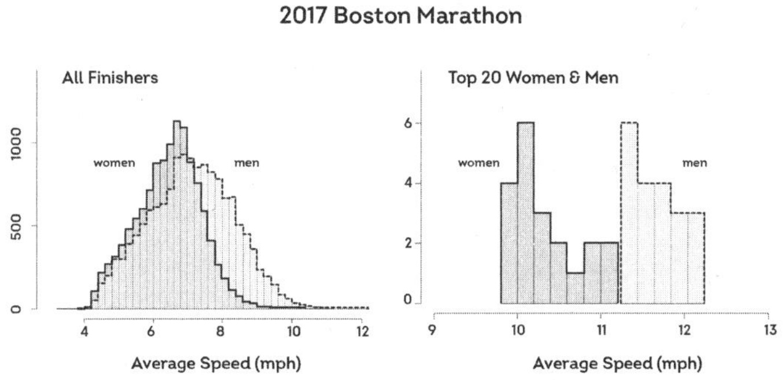

Figure 7.3 Average running speeds (miles per hour) for finishers in the women's and men's categories in the 2017 Boston Marathon.

SEX ON THE BRAIN

The effects of sex hormones on the brain are far more controversial. There are small, but consistent differences (on average) in the detailed anatomy of male and female brains. These differences must result from the effects of sex hormones or gendered experiences (or both), since there are no "male" and "female" genes for brain development. Still, it's unclear what, if anything, these minor differences mean for cognition or behavior.

Studies have looked for sex differences in intelligence and cognitive performance for decades, since the dawn of intelligence testing. No careful studies have shown any sex differences in average IQ, but some patterns do emerge. For example, males tend to do a bit better in spatial reasoning (like mental rotation of a 3D object), and to be more aggressive, while females tend to show more interest in people rather than things. But average differences are small, and there is a lot of overlap, even more than we see in physical size and strength—you would do a very poor job predicting verbal or spatial abilities based solely on a person's sex. But we see these small differences in average scores consistently across studies.

The potential impacts of sex hormones on brains and behavior raise a number of big questions. Could testosterone and estrogen, with their many downstream effects throughout the body, influence brain development in some way that affects cognition? Or are the apparent trends in ability and behavior completely explained by the way males and females are raised and socialized? What about broader, and more personal, aspects of the way we think and feel? Do sex hormones turn males into boys and men, females into girls and women?

As we grapple with these questions, the distinction between sex and gender is crucial to keep in mind. Sex refers to chromosomes, anatomy, hormones, and other measurables that distinguish male and female bodies (and occasionally don't, in the case of intersex individuals). Gender refers to the way that society ascribes masculinity or femininity to various people, objects, concepts, or behaviors. Gender identity is the way a person feels about themselves. Gender is cultural and subjective, difficult to measure, and can even apply to things that don't have a sex. Colors are sexless, and you can't measure the gender of pink or blue, but if you grew up in the U.S. you know intrinsically which is a "girl color" and which is for boys.* Conversely, most plants and animals have sex, but only humans have a gender identity. There are male and female ginkgo trees and great white sharks, but as far as we

---

* If you answered "Yes, of course. Boys usually wear pink dresses," your gender IQ is about a century behind the times. Pink dresses were gender-typical boys' clothes in the late 1800s.

can tell they lack the self-awareness and culture necessary for gender identity.

We tend to take it for granted, but it is remarkable how reliably sex predicts a person's gender identity. Large-scale national surveys in the U.S. by the Williams Institute at the University of California Los Angeles (UCLA), a research group dedicated to understanding issues facing the lesbian, gay, bisexual, transgender, and queer (LGBTQ) communities, consistently finds that more than 99 percent of adults identify as the gender aligned with their sex (cisgender). Males identify as men, females identify as women. That rate falls just slightly, to about 98 percent, among younger individuals, ages thirteen to twenty-four. Fewer than 2 percent of adults at any age identify as transgender or nonbinary.

Sex predicts sexual orientation as well, to almost the same degree. Most people develop an attraction to the opposite sex. The Williams Institute puts the percentage of U.S. adults who identify as either gay or lesbian at about 4 percent, with similar numbers in other Western countries. The numbers are comparable for men and women, but change a bit for both when the question is asked in different ways. Around 8 percent of U.S. adults report having had a same-sex sexual encounter, and around 11 percent report at least occasional attraction to same-sex individuals. The percentage is also greater in younger generations. Still, across surveys and age-groups, over 90 percent of adults report being sexually attracted to opposite-sex individuals (although not necessarily exclusively).

Is the strong correspondence between sex, gender, and sexual orientation due to nature or nurture? There's a school of thought in psychology and sociology that sex differences in brains and behavior, including gender and sexual orientation, are purely products of our environment, the socialization we experience from infancy: boys being taught to be boys and girls taught to be girls.

This perspective was part of the feminist response to the biological determinism that prevailed well into the mid-1900s. Biological determinists viewed the strong correspondence of sex and gender as evidence that all aspects of gender, including traditional gender roles and limits for women, were an inescapable outcome of innate and immutable differences in the way that male and female bodies function. To them, sex and gender were inseparable. These are the folks who argued, as late as the 1960s and '70s, that women shouldn't be airline pilots or run in the Boston Marathon.

Feminist scholars fought back, pointing to the clear evidence of socialization on gendered behaviors and the misogyny embedded in biological determinist arguments. In doing so, they argued sex was separate from gender, a distinction many of us still find useful today. Women's rights and opportunities in school, sports, and professional life expanded, and women's excellence proved their traditionalist critics wrong. The prevailing view in the humanities changed, skeptical of *any* biological influence on gender or sex differences in behavior. It seemed possible that the tight correlations between sex, gender, and sexual orientation were due to socialization, the way we're raised.

Research in the biological sciences has made it increasingly difficult to ignore the contribution of sex hormones, however. As Carole Hooven reviews in her book *T: The Story of Testosterone, the Hormone that Dominates and Divides Us*, sex hormones clearly affect the brain in other mammals. It would be surprising if they had no effect in us.

We know, for instance, that sex hormones can affect our brains and the way we feel. Fluctuations in estrogen throughout the menstrual cycle have been linked to mood changes, as has the loss of estrogen with menopause. Low testosterone levels in males have been implicated in depression. Testosterone also shapes the way both men and women experience winning and losing. Winners, whether it's a chess game or a soccer match, have elevated testosterone levels and jubilation, while losers experience a testosterone crash and dejection. You don't even need to be a participant—spectators experience the winner effect as well. Sex hormones get into the brain.

It's plausible that sex differences in hormone levels influence the small differences we find in cognition or behavior, like the male advantage in spatial manipulation, but the evidence at the moment is lacking. It's not easy to design studies that isolate the effects of hormones, particularly when the critical period of exposure might be in the womb or early infancy, or might develop over years. Demonstrating the effects of sex hormones on gender identity and sexual orientation is even harder, not to mention more controversial. For (hopefully obvious) ethical reasons, we don't do the same experimental studies in humans that we can do with rodents, manipulating hormone and environmental influences independently of sex. In the real world, as we develop through childhood and adolescence, sex and socialization are usually aligned. Males are raised as boys, females are raised as girls, making it impossible to distinguish their impact.

There are some people for whom sex and socialization are *not* aligned, and their experiences say a lot about the interplay of nature and nurture. In rare cases, chromosomal males (XY) with normal testosterone production and receptors do not form a penis during development, or their penis is injured so badly soon after birth that it is removed. In the past, surgery was often performed in these cases to make the external anatomy female and these individuals were raised as girls, their testes removed to prevent further testosterone production. If gender was entirely a product of socialization, we'd expect these individuals to grow up identifying as girls, and many of them did. Yet, by late childhood or adolescence, many (perhaps 30 to 50 percent) identified as boys or transitioned to live as boys—far more than the rate of transgender identity in the general population. Such high rates of gender transition are a major reason that medical guidance in these cases has shifted away from surgical intervention and female gender assignment.

People with male chromosomes and the condition 5-ARD provide another example of hormonal influence on gender identity. These individuals, as we mentioned above, produce testosterone, but cannot synthesize a related androgen, DHT, and are often born with female external anatomy. Around half of chromosomal males with 5-ARD who are raised as girls go on to identify as boys.

In both cases, individuals who come to identify as men also report sexual attraction to women, suggesting sexual orientation might also be influenced by testosterone exposures in the womb. Work by David Puts, a researcher at Penn State, supports this idea. Chromosomal males with 5-ARD, androgen insensitivity, and other intersex development differ in the amount of testosterone they produce and their cells' ability to sense it. Puts's analyses indicate that those with more testosterone and greater sensitivity are more likely to develop sexual attraction to women in adulthood.

We see something similar among chromosomal (XX) females exposed to high levels of androgens due to congenital adrenal hyperplasia. These individuals are typically raised as girls, and the overproduction of androgens is stopped in infancy with medication. These girls often show a preference for toys and games that are usually preferred by boys, and there is some evidence that they are more likely to identify as boys than are girls without congenital adrenal hyperplasia. They're also more likely to pursue "masculine" careers, like mechanics or carpenters. As adults, women with congenital adrenal hyperplasia have a greater likelihood of being attracted to other women than we find in the unaffected population.

These experiences suggest that sex hormones affect our brains in ways that shape our gender and sexual orientation, and that these effects start early, in the womb. That *doesn't* mean socialization and environment aren't important. Many of the people affected by differences of sexual development and other rare conditions live their lives in the gender they're raised, not the one aligned with their chromosomal sex, gonads, or hormones.

We see these complexities echoed in the broader population. Millions of transgender people identify as a gender that is not aligned with their sex chromosomes, hormones, or anatomy. In most, if not all, cases, their gender identity doesn't align with the way they were socialized as children, either. The same can be said for gay and lesbian adults. Their sexual orientation doesn't align with what we might expect from sex or socialization. Neither, by itself, gives a satisfactory account of these deeply personal traits.

Adding in genetics doesn't help. We haven't mentioned genes much in our discussion of gender and sexual orientation because there is very little to say. Twin and sibling studies have suggested that transgender identity and same-sex attraction are modestly heritable, but as we've discussed in previous chapters these studies tend to overestimate genetic effects and underestimate environmental influences. No studies to date have identified any genes associated with transgender identity. GWAS studies have identified a handful of alleles that are weakly associated with same-sex attraction and behavior, but that's about it. You can't predict a person's sexual behavior or identity from their genes.

Ultimately, our efforts to understand the complexities of identity and behavior run aground on the same rocky shores as our efforts to understand consciousness and the mind. We know they are products of the brain, and therefore fundamentally a part of our biology, but we don't really know how they work—not at a level that allows us to compute the many contributing factors (hormones, genes, social influence, nutrition, stress, neurotransmitters . . .) and make reliable predictions about our interior life. All we can say, with the evidence at hand, is that biology and socialization both matter, and that neither on its own determines how we think or who we become.

EXISTENTIAL DEFINITIONS

With all the hand-wringing from cultural traditionalists in the U.S. and Europe over recent shifts in societal norms around sex, gender, and sexual orientation, it's notable that a number of traditional societies have long wrestled with these issues and developed a nuanced cultural understanding. There are more than two dozen farming, foraging, and fishing cultures worldwide with a Story of How the World Works that recognizes and embraces gender identities and sexual orientations beyond the traditional Western strictures. If Hadza men can figure out the chemistry to make poison that can kill a giraffe, we shouldn't be surprised by the sophistication of our farming and foraging neighbors. Perhaps they can teach us something.

Current debate in the U.S. and other Western societies centers around a seemingly simple question: What is a woman? It's a reasonable question, and one that a book about the human body should be able to answer. The fact that it's become a flash point in the culture wars tells you it's a deceptively complex and incendiary issue.

First things first. Words mean what we, as a community, agree they mean. There's no "right" answer, handed down from above or discernible from a mass spectrometer. Common usage determines dictionary definitions, not the other way around. And definitions change all the time. A "computer" used to be a person who did math for a living, now it's a machine. Definitions also differ by context. Is a hot dog a sandwich? In some legalistic sense, maybe, but you don't ask for a sandwich from a hot dog stand. Is a tomato a fruit? To a botanist, yes, but not to a chef.

The traditional definition of a woman, preferred by many on the conservative right, would be something like "an adult human female." That's similar to what you'll find in the Merriam-Webster dictionary, reflecting the way most native English speakers use the word. It's the definition I grew up with. Society has developed that meaning for the word because it works. Even today, in the midst of a cultural reckoning on gender, the traditional definition works well over 98 percent of the time. But as clear as it seems, and despite what the biological determinists might claim, there's ambiguity in the traditional definition of a woman as an adult human female.

We first need to know what an "adult" is. Are we talking legal definitions? If so, which one? The age of consent for marriage (fifteen in Hawaii, eighteen in Kentucky)? Driving a car (sixteen in most states)? Voting (eighteen)? Drinking (twenty-one)? Or are we talking about biological adulthood? That's a highly variable process with any number of milestones we could use. Is a girl who gets her period at nine years old a woman? How about when her skeletal growth plates fuse (and which ones)? Or do we wait until her wisdom teeth come in?

Leaving the definition of "human" aside for the moment (could a Neanderthal be a woman?), we'd also need to agree on what a "female" is. Are we talking about chromosomes? Gonads? External anatomy? Hormones? As we've seen, you'll include different groups of people depending on the criteria you use.

I suspect that for most people, these ambiguities are just another instance of language's imprecision. The traditional definition usually works well enough in conversation, and additional clarification is always possible if necessary. Using "woman" as a shorthand for "adult female" isn't a political statement that's meant to exclude. But for others, particularly on the political right, ignoring intersex people and excluding transgender women is the point. They reserve the word "woman," along with its cultural weight, for people with typical female bodies.

An alternative definition, favored by many on the progressive left, is that a woman is anyone who identifies as a woman. Technically, it's still a definition based on anatomy—the brain, after all, is where identity must reside. But the draw is that this definition favors identity, that ineffable *me*-ness we all feel in our core, over any measurable aspect of our biology. In doing so, it bypasses the arguments and ambiguities surrounding chromosomes, anatomy, and hormones. It's a definition that's inclusive of transgender women and intersex people who identify as women. It gives these marginalized and vulnerable communities a home.

In its fullest expression, an identity-based definition of gender makes biology irrelevant. A person's sex might influence their identity, but identity by itself is sufficient to establish gender. Trans women are women because they identify as women, full stop.

A more nuanced argument for identity-based gender definitions is that biological sex is continuous, not categorical, and therefore can't be used to place people into gender categories. Proponents of this view, like biologist Anne Fausto-Sterling and biological anthropologist Agustín Fuentes, highlight the bodies and experiences of intersex people and the shared foundation of male and female development. In our Pacific island metaphor, they focus on the isthmus connecting the male and female peaks and the massive seamount that they share. In their view, these connections mean that sex isn't truly binary but, instead, a continuous spectrum. (Fausto-Sterling famously proposed dividing humans into *five* sexes—not as a serious attempt at classification, but as a statement about the arbitrariness of the male-female dichotomy.) And while they acknowledge that male and female bodies differ in some important ways, they emphasize the similarities in body or behavior. Sex and its connection to gender is so complex and fluid, they argue, and the harm of excluding transgender and intersex people so great, that we should honor a person's identity. Trans women are women because they identify as women, full stop.

Others have pushed back against identity-based definitions of gender, arguing that having a female body is inextricably entwined with being a woman. Traditionalists on the political right make this case, of course, but so have a number of prominent feminist scholars on the progressive left, including women, like the writer Chimamanda Ngozi Adichie, with a history of championing LGBTQ rights. Their claim is not that biology determines gender in a mechanistic way, but that the experience of living in a female body—both the visceral, internal experience and the way one is treated by society—is integral to being a woman.

It's anyone's guess where the definition of "woman" will wind up. Absolutist views on either side would seem to face an uphill battle in the court of public opinion, and it is cultural consensus that determines what words mean. Ignoring intersex people and refusing to acknowledge a person's lived identity seems callous and harmful. Insisting that biology doesn't matter requires people to ignore obvious differences in bodies, both theirs and others'. Some people may be happy to overlook the occasional penis, but others might not be so sanguine.

Perhaps we'll adopt the approach of other cultures, those with a long history of acknowledging and even revering nonbinary and transgender people. In many of those cultures, the word for "woman" refers to adult females (*fafine* in Samoa, for example), while a separate word is used for transgender women (*fa'afafine*).

Crucially, the distinction isn't loaded with negative connotations because these cultures embrace intersex and transgender people. Or perhaps we will learn to live with ambiguity. Like "adult," the legal, cultural, and moral dimensions of "woman" might always depend on the situation. Language is full of categories with flexible boundaries, from sandwiches to fruit, and there's no reason to expect that the words we use for ourselves will be different.

CLASHES OVER THE WORD "WOMAN" ARE SO FIERCE BECAUSE THEY are part of the larger debate around transgender inclusion. Women's-only institutions, from colleges to rape crisis centers, are being asked to open their doors to people with male bodies, or to justify their exclusion. If an institution's mission is to serve women, the definition is existential. Most U.S. women's colleges, for instance, have revised their policies in recent years to admit trans women. Women's health services, which specialize in caring for people with female bodies, might see things differently, as might women's crisis centers that care for women victimized by people with male bodies.

For other institutions, the definition of a woman is secondary to questions about the safety and fairness of including people with male bodies. Women's athletics is the classic example. In these cases, understanding our biology can help ground the debate.

Most sports put a premium on size, strength, or endurance, traits with a clear male advantage that isn't erased with testosterone-blocking drugs. The difference is more pronounced at higher levels of competition. Take running, for instance. In the 2017 Boston Marathon, nearly all of the top one hundred male finishers were faster than the best one hundred females. Performance was completely binary among the very fastest runners, with no overlap at all between the top twenty males and top twenty females.

It's not just marathons. For shorter races, from 100 meters to the 10K, the current women's world records are slower than the Olympic-qualifying times for men. No woman has ever run a sub-four-minute mile. The current female record, 4:07.9, is about as fast as the male record in 1933 and slower than the current record for fifteen-year-old boys. Allowing trans women and others with male bodies to race in the women's category would promote inclusion, but at a substantial cost to fairness for athletes with female bodies.

Analyses like these are already being used in international sport. A study by World Rugby found the sex difference in size and speed presented an unacceptable safety risk to including trans women in women's rugby, outweighing the benefits of inclusion. On the other hand, a study of archery and other shooting sports found that male bodies don't confer an obvious advantage. Men's and women's records differ by just 1 to 2 percent on average, and in some disciplines women's records are better. The authors of that report conclude that integrating trans women into shooting sports would satisfy the goal of inclusion with negligible impact on fairness.

There's tantalizing evidence that there could be a *female* advantage in ultramarathons, open-water distance swimming, and other ultra-endurance races. The numbers of competitors in these events are small, and men still typically outpace women, particularly in well-attended races that provide larger samples of male and female competitors. But if more research were to demonstrate a female advantage, ultra competitions would presumably need to weigh the fairness of allowing transgender men to compete in the men's category.

These sorts of analyses still leave us to consider complex cases involving intersex people whose bodies don't fit the usual male-female distinctions. Take Caster Semenya, for instance, an

Olympic gold medalist middle-distance runner with the condition 5-ARD. People with 5-ARD have Y chromosomes and produce testosterone in the normal male range, and they have functioning testosterone receptors, conferring advantages in strength and endurance. Still, people with 5-ARD often have external anatomy that looks female, and might not know anything is different until puberty, when their development diverges from their female peers. Caster grew up as a girl, identifies as a woman, and has competed her whole career in the female category. For the past decade, Caster has had to fight to compete, as rules of eligibility have changed. Intersex athletes like her present real, complex challenges in developing rules that emphasize fairness (the top priority for most sporting federations) while still respecting athletes' rights and dignity.

Difficult discussions like these, in athletics and in other contexts, will be fairer and more transparent if we're clear-eyed about the ways that male and female bodies differ and the ways that they're the same. Focusing on our biology might even offer a patch of common ground to traditionalists and progressives. Like our imagined Pacific island, the degree of separation between male and female bodies depends on the trait of interest. In some respects, sex *is* effectively binary, while in many others there's no meaningful difference at all. As the culture around sex and gender matures, the definition of a woman—or at least, the rules about who is allowed in women's communities—might grow to depend on the context as well.

ON THE OTHER HAND

Integrating a modern understanding of human biology into the culture around sex and gender has meant a rewrite, or at least a revision, to our collective Story of How the World Works.

Traditionalists have been vocal in their resistance, and it's unclear at the moment where and how the dust will settle. Your guess is as good as mine. But it turns out we have a helpful example of how this new story might play out. It's the Story of Being Left-Handed.

Handedness is similar to gender and sexual orientation in some important ways. To begin with, we don't really understand how handedness develops—neither genes nor environment reliably predict it. Something in the way our brains grow produces right-handedness in nearly everyone, upward of 90 percent of the population. But some of us, myself included, develop differently. About 10 percent of us are lefties. Another 2 percent don't identify as either. Within handedness groups there are subgroups. I write lefty but throw righty, while most people prefer the same hand for all tasks. Handedness emerges early, usually by age two, and might even be detectable in the womb.

Across much of the world, the old Story of Being Left-Handed held that it was deeply unnatural, reflecting some failure in the normal course of development. In the Bible, the good people are on God's right hand, not the left. In Muslim culture, the left hand is unclean. Such left-phobic sentiments were enshrined in language. *Sinister*, the Latin word for "left," became a synonym for evil. The English "left" derives from *lyft*, meaning "broken." Up through the twentieth century, kids in the U.S. and elsewhere were heavily socialized to be right-handed, and corrected by force if necessary. Only 3 percent of kids born between 1900 and 1910 grew up to be left-handed adults.

Then, for reasons that aren't entirely clear, the Story of Being Left-Handed changed. In the U.S. and Europe, the stigma over left-handedness lifted. It went from being pathological to just being different. Socialization didn't disappear. Scissors, guitars, school desks, and the rest are still predominantly made for righties. We shake right hands, put our right hands over our hearts, raise our right hands to swear an oath. We haven't changed our language—it's still good to be right and righteous. But as someone who grew up a lefty, I can tell you it wasn't a big deal. About 10 percent of Americans born in the 1970s like me grew up left-handed, and the rate has held steady ever since.

Changing the Story of Being Left-Handed meant more kids were able to grow up being themselves. It meant no more attempts to convert people to being right-handed, which is often ineffective and can cause real harm, including stuttering and learning disability. Changing the story didn't require us to abandon handedness as a binary concept. We know there are ambidextrous people, and we appreciate them all the more because we understand that they are rare. We came to embrace handedness (when we think about it at all) as a mixture of biological and social influences, free from the weight of being good or bad.

We're witnessing similar changes in the way society thinks about sexual orientation and gender identity. Until the mid-1970s, homosexuality was officially classified as a mental disorder by the American Psychiatric Association. Until 2011, being openly

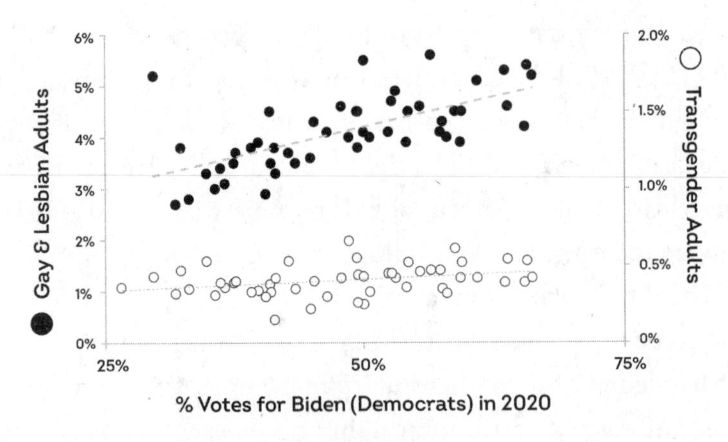

Figure 7.4 Percentage of gay and lesbian adults (filled circles) and transgender adults (open circles) in each U.S. state plotted against the share of the 2020 presidential vote for Joe Biden, the Democratic candidate.

gay was grounds for dismissal from the military. But by 2015, same-sex marriage was legal in all fifty states. As attitudes around homosexuality shifted, the proportion of the population identifying as gay or lesbian steadily increased. In a large 2019 study of sexual orientation using the UK Biobank sample of British adults, the proportion identifying as gay or lesbian increased from around 2 percent of those born in 1940 to over 6 percent of those born in 1970.

We can see some of these dynamics playing out today in the U.S. The percentage of people who identify as gay or lesbian is greater in less conservative states (see figure 7.4). There's a similar trend for transgender people. These comparisons across states highlight both the power and limitations of socialization. The most liberal states have a greater proportion of openly gay, lesbian, and transgender people, but the impact is modest—they are part of the community in every state. Reducing the stigma around sexual orientation and gender identity doesn't open the floodgates and make everyone gay or transgender, as some conservatives seem to fear, any more than letting kids use their left hands turns everyone into lefties. All it does is allow more people to live authentically.

Making our social environments more livable for everyone is an important goal. Our protein robots don't work as well or last as long under a constant barrage of stress. The world out there is tough enough already, and it's only heating up.

8
ENVIRONMENTAL PROTECTION

One of the first things you learn if you study ecology is the beneficent nature of the sun. Life needs energy, and the light from our star flows through the living world, fueling and sustaining the earth's ecosystems from the lowliest slime molds up through the most majestic beasts. "We are made of star stuff," as Carl Sagan noted so poetically, our atoms forged inside the bellies of old, exploded stars. Your body, and indeed your life, is also temporary storage for the sun's energy. Captured by photosynthetic plant cells, sunlight was used to assemble carbon, water, and other molecules into living tissues. You ate those tissues and their embodied sun glow, or you ate animals that did, and you constructed your own protein robot from the building blocks. The energy your muscles, nerves, and other cells use to move you around and ignite your thoughts came from the fusion reactor deep within the sun. We are solar powered.

That's not how it feels, though, when you're traipsing about the savanna under a blazing equatorial sun. It feels like you're walking around inside an air fryer. When I'm working with the Hadza or the Daasanach communities, I'm always shocked at how intense the heat is. No matter how hard I try to remember the last visit or steel myself for the blistering air, I spend the first few days feeling like the sun is right behind me, breathing down my neck, pressing up against me. It's visceral and unsettling, like the horror movie scene where the victim realizes the axe murderer is already in the house. *The sun is trying to kill me.*

Hadza, Daasanach, and anyone else who's grown up in heat like that learn to get out of harm's way. No one in those communities stands around idly in the sun, not even for a minute. Everyone hides beneath the cover of the trees, tucking themselves into the shade, like hardened soldiers evading enemy aircraft. And while they aren't as wimpy about it as I am, they know to take the heat seriously. On a landscape full of poisonous snakes, sneaky hyenas, and hungry lions, you're more likely to die from dehydration or overheating than you are to end up as dinner. Lions only manage to pick off a hundred or so people each year across all of Africa. Excessive heat kills tens of thousands.

Thankfully, our body is equipped with an array of tools to help us fight the elements. We wear these adaptations like a suit of armor. From the sweat dripping from our pores to the blue lips we get standing around in the cold to our particular shade of pink or brown, our skin protects us from the elements and keeps our inner world steady and stable. Beneath the surface, our kidneys work to maintain the precise degree of hydration, all the while ridding us of any toxins that find their way inside. Along with cultural adaptations like fire, clothes, and shelter, these clever adaptations allow us to thrive in every habitat and climate on the planet.

The elements and our adaptations to them have also come to

divide us. Throughout history, no trait has been more divisive than the color of our skin. Today, no challenge is more polarizing than the warming of our planet. At work, no one is more annoying than the colleague who insists on chugging a gallon of water every day and telling us about his current cleanse. Like the hidden infrastructure and essential workers keeping our environment clean and safe, our body's department of environmental protection is overlooked and misunderstood.

CHECK YOUR TEMPERATURE

Some 230 million years ago, in the Late Triassic period, a group of shrew-sized animals called synapsids began shaking things up. They were getting out of the reptile family business, trying out new ways of making a living. Their skin changed, ditching a dry, scaly hide for something smoother, full of oil-oozing glands that opened out into pores. Cells that built scales evolved to produce hair. They radically changed the way they reproduced, giving birth to live young and feeding them a calorie-rich liquid produced by specialized glands in their skin. These were the earliest mammals.

This paradigm shift in vertebrate evolution was fueled by a wholesale change in metabolism. Your body is a massive chemistry set, with trillions of reactions happening every second to break molecules down, build them back up, move them through membranes, and more. Collectively, all that cellular work is your metabolism. Metabolic chemistry happens faster when it's warm, and the reactions can be finely calibrated if the temperature is stable. Most animals are at the mercy of the environment, relying on warmth from the sun to get their chemistry cooking. When they get cold, their metabolism slows down. That's why snakes and other reptiles sunbathe on warm rocks. Early mammals evolved an inner furnace, running their metabolism so fast that the heat it generated kept their bodies at nearly 100 degrees Fahrenheit (38°C), even when the air temperature dropped.

A faster metabolism meant more of everything. Faster running for longer distances, more activity in the cool of the night, faster growth and reproduction. It was such a successful strategy that birds would independently evolve it millions of years later. But it came with a cost. Reptile body temperatures can fluctuate by twenty degrees or more without any problem. Mammalian metabolism evolved to be so well tuned to their particular body temperatures that the whole system would fall apart if body temperature deviated by even a small amount. Cold and heat could be lethal.

With body temperature literally a matter of life and death, early mammals evolved a suite of tricks to keep their insides comfortable, adaptations we find in our own bodies today. You've got specialized nerve cells embedded in your skin and neurons in the hypothalamus (is there anything the hypothalamus *can't* do?) to track external and internal temperatures. When temperatures deviate even a little from your body's preferred temperature setting, your heating or cooling systems kick in to nudge the system back in line.

Average human body temperature is 98.6 degrees Fahrenheit (37°C), but people vary by a degree or two in their personal setting. Some of us run hotter or colder than others, as anyone who has tried to share a bed and blankets (while sleeping) with another human can probably attest. Some of this variability is tied in obvious ways to metabolic rate and body size, which influence how much heat you generate and how insulated your body is (fat is a wonderful insulator). People with hypothyroidism, for example, have a slower metabolic rate and run cooler. If you're larger, you'll tend to run warmer. Men also run a bit warmer than women, most likely because men tend to carry proportionally more heat-generating muscle and other organs, and proportionally less fat. Your body runs coolest in the early morning, warmest in the late afternoon, consistent with the circadian rhythm in metabolic rate.

Dieting can lower your temperature, because the body responds to calorie cutting and weight loss by reducing energy expenditure. Older adults, past sixty or so, tend to have lower body temperatures as well, which corresponds to the slow decline in daily energy expenditure in later years. And, of course, your environment affects your temperature. When it's really hot your temperature can rise a little, and it can drop a bit when it's really cold, but we also adjust to longer-term, seasonal changes in temperature. Our bodies tend to run a bit cooler in the summer and warmer in the winter, trying to stay a step ahead in dealing with the outside temperature.

Even when we account for all these factors, there's still a lot of variability in individual body temperatures that we don't understand. Intriguingly, there's some indication that body temperature could predict your risk of dying. A recent study of body temperature in over thirty-five thousand adults found that running just two-tenths of a degree warmer was associated with an 8 percent increased risk of dying within a year. We're talking about *relative* risk here, so the impact on the *absolute* risk of death is still pretty small (see chapter 4), but it's an interesting finding. Strangely, normal human body temperature has fallen over the past century by about one degree Fahrenheit (0.5°C). This unexplained cooling is likely related to the small decline in basal metabolic rate that has also been observed since the early 1900s. I suspect both are due to the lower pathogen burden we experience in our modern antiseptic environments. Others have argued diet changes could be a factor. No one really knows.

Wherever your personal internal thermostat is set, when you're cold, your body works to generate more heat and to trap it inside, particularly around the vital organs. You might get goose bumps ("piloerection" is the scientific term), an evolutionary holdover from the days when our more apelike ancestors still had fur. Those tiny bumps are the muscles around your hair follicles flexing, lifting the hairs erect and (if you have *a lot* of body hair) creating a little layer of insulating dead space right above the skin. You also get goose bumps when you're frightened, because lifting all your hairs makes you look larger and less vulnerable—or it would, if you still had fur.

Other responses to cold are more effective. The blood vessels near the skin's surface and in your arms and legs constrict, reducing the flow of warm blood to the cool skin and thereby reducing the heat lost to the environment. It also takes some of the color from your appearance, with less red blood just below the skin's surface, leaving you pale and with blueish lips.

You also generate more heat, ramping up your metabolic rate in two main ways. Little pockets of brown fat, specialized tissue built just for making extra heat to keep you warm, switch on. These small deposits lie mostly in the base of your neck and upper chest, and their mitochondria burn calories just to create heat, rather than making ATP. You'll also shiver, if your internal temperature drops low enough, which is your muscles contracting and relaxing rapidly. Both of these responses are involuntary. Your body doesn't leave the important work of not dying up to personal choice. They are also adaptable. The more you're exposed to cold temperatures, the quicker you respond and the less uncomfortable you feel.

We've got long-term solutions to cold as well, both biological and cultural. Populations that live in cold climates for generations tend to be a bit heavier and stockier, body proportions that help keep heat in. The cold is enough of a threat to survival, not to mention the energy it can steal from growth and reproduction, that alleles associated with larger bodies and shorter arms and legs are favored. This is the mirror image of the tall, thin body proportions we saw with the Daasanach in chapter 2.

And, of course, the clever primates we are, we've got an impressive set of cultural adaptations to cold. Fire is the obvious one, and it's well over a million years old. There's a direct and unbroken chain of invention and development from the first Paleolithic fires to the furnace that heats your home. Shelters and clothing are the other big cultural solutions to cold, but their origins are less clear. They don't preserve well in the archaeological record. But Wonderwerk Cave in South Africa, site of the earliest well-preserved fire, is about 1 million years old, evidence that our genus, *Homo*, was smart enough by then to come in out of the cold when the temperatures dropped.

Our modern solutions to staying warm are so effective that they've taken our bodies largely out of the picture. Xueying Zhang recently led a large collaborative research study, along with myself and other scientists who specialize in energy metabolism, measuring total daily energy expenditure—all the calories burned each day—in over 3,200 American adults. We found no effect of outside temperature on the energy burned each day. People in the cold parts of the country, even during the winter months, didn't burn more calories than those in warm environments. Our clothes and buildings are so effective that people rarely shiver.

By comparison, traditional pastoralists living with their herds in Tuva, Siberia, burn around 20 percent more energy each day during the cold winter months when outdoor temperatures hover around minus sixteen degrees Fahrenheit (–27°C) and even the temperatures in their homes fall below freezing at night. Interestingly, Tuvan adults also appear to vasoconstrict *less* than other populations when exposed to cold, which keeps their hands warmer and more functional under cold conditions. Whether that difference is due to growing up and living in cold environments or to a local genetic adaptation in their population has yet to be determined.

The body works to keep itself warm because the costs of getting cold are dire. A drop in core body temperature of just four degrees Fahrenheit (2°C) impacts organ function so severely that the body usually can't recover on its own. Nearly one thousand people in the U.S. die each year from hypothermia.

Remarkably, it *is* possible to survive a severe drop in body temperature during cold-water drowning. It's not entirely understood how some people survive extended periods below the surface (and to be clear, most don't), but it seems icy water can cool the body so rapidly, and slow metabolism so much, that the person enters a sort of suspended animation that protects them from the lethal effects of oxygen deprivation. It's most often seen with children. In 1986, a two-and-a-half-year-old girl in Utah was plucked from an icy creek after more than an hour underwater. She was limp and blue, her heart completely still, and her body temperature had dropped to sixty-six degrees Fahrenheit (19°C). After two hours of heroic medical effort and careful rewarming, her heart was beating normally again. A year later, she was walking, talking, and playing like any normal kid her age. The oldest-known cold-water near-drowning survivor was a sixty-two-year-old Canadian construction worker whose tractor broke through early-spring ice and plunged him into the Red River for fifteen minutes.

HEAT IS LESS FORGIVING. JUST A COUPLE OF DEGREES ABOVE NORMAL body temperature can cause heat exhaustion (fatigue, dizziness) and trigger seizures in people who are vulnerable. Your body temperature can rise up to 105 degrees Fahrenheit (41°C) if you're exercising in the heat or in the throes of a major fever. It's uncomfortable (and to be avoided if possible), but survivable. Much above that and you're in real trouble, risking damage to the heart, brain, and kidneys, or even death, from heatstroke.

Scientists and activists have been banging the alarm bells for decades now about the dangers of heat on our warming planet. Most of the doomsday scenarios from climate change involve freak storms, rising seas, water shortages, and climate refugee crises—and those should absolutely scare you if they don't already. But as Jeff Goodell explains in his book *The Heat Will Kill You First*, there is also a real danger from the heat itself. People working outdoors, like the farmworkers who put food on your table, are most vulnerable. But the heat is coming for all of us. The human body can't cope with air temperatures above ninety-seven degrees Fahrenheit (36°C) when the relative humidity hits 66 percent, as it often does here where I live in North Carolina. In a dry heat, with relative humidity well below 50 percent, like in Phoenix, Arizona, the maximum air temperature you can handle is about 105 degrees Fahrenheit (41°C). Those temperatures used to be exceptional, but are increasingly common. Summer heat waves regularly kill tens of thousands of people across the U.S. and Europe each year.

The modern solution to a warming planet is to crank up the air-conditioning. Effective indoor cooling has made the deserts habitable, with cities like Las Vegas and Dubai growing like hermetically sealed Martian outposts of glass and steel. The irony is that all that cold air relies on a steady supply of energy from fossil fuels to run cooling units that generate CO_2 and heat—more heat than they remove from the air indoors. Air-conditioning cools our homes, but warms the planet.

The main reason heat is so deadly is that our bodies have a limited number of tricks to cool down. You are constantly generating heat to stay alive—more if you're working or exercising—and you need to dump that heat into the environment so it doesn't build up to lethal levels. That's much harder when the air is already warm. You can dilate the blood vessels under your skin and in your extremities, pushing more hot blood to the surface. You can stop moving to reduce the energy you're burning and the heat you generate, and your brain does a reasonably good job making you feel exhausted and sluggish to slow you down when it's hot. (Even professional athletes competing for the podium run slower in the heat.) And over many generations, we sometimes see selection for tall, thin body proportions like we find with the Daasanach, which increase the ratio of surface area to body mass and make it easier to get rid of excess heat.

The most effective trait you've got for keeping cool is also the most radical: your naked, sweaty skin. Unlike every other primate on the planet, we have evolved ultrafine body hair that leaves us effectively hairless. We've also got the sweatiest skin on the planet, with more sweat glands than any other mammal and over ten times more than our ape relatives. You can easily produce one liter of sweat per hour, the cooling action of evaporation dumping hundreds of kilocalories' worth of heat into the air. Your sweat glands are also clever, adjusting to produce more sweat and cool you more aggressively after just a few days of exposure to heat.

Sweat doesn't fossilize, but it seems most likely that our strange damp, hairless skin evolved around 2 million years ago as early members of our genus, *Homo*, were adapting to an active, hunting-and-gathering lifestyle. These populations lived in equatorial Africa, and sweating would have allowed them to stay active and cover longer distances in the heat of the day. If they were running down antelope and other big game, as Dan Lieberman and others have argued (see chapter 6), sweating would have provided a killer advantage, allowing Paleolithic *Homo* to stay relatively cool while they ran their prey into heat exhaustion. We likely retained the hair on our heads as an adaptation to keep the hot sun off our scalps and prevent it from cooking our brains. (Eyebrows and facial hair seem to be all about social communication and attracting mates.)

The only downside to sweating, aside from that ripe locker room aroma, is all the water you need. Your internal chemistry set is water-based, and you need to keep it that way. Running out of water is just as deadly as overheating.

BLOOD IS CLEANER THAN WATER

Life and all of its chemistry evolved in the seas. When the first vertebrates started making a living on land around 370 million years ago, they needed to figure out a way to take their water-based chemistry with them. Land animals became walking reservoirs, taking the sea along like a personal life-support system. Skin changed from damp to dry to reduce evaporation. Kidneys evolved to conserve water. The blood you've got sloshing around your body today is a small, salty ocean, an evolutionary legacy of your marine ancestors.

The basic strategy for maintaining your personal water-based life-support system in a dry, terrestrial world is to ingest water and limit how much you lose. In wet habitats, that's easy. Animals that live in or near the water, like beavers, drink whenever they like and don't need to worry much about how much they lose each day. Our ape relatives, living in rainforests, have it easy as well. Chimpanzees, orangutans, and gorillas consume so much water in the juicy fruits and leaves they eat that they rarely need to drink. They can happily go days or even weeks without drinking.

Humans have it harder. We live in habitats all over the globe, including dry savannas and deserts. And we cook our food, which often dehydrates it. If we don't drink, we're dead. Even with our kidneys doing their best to minimize water loss, we can't go more than two or three days without drinking or we'll die.

My colleagues and I investigated ape and human water use in a recent study and made a fun discovery. We used safe, stable isotopes of hydrogen and oxygen to track the flow of water through the body in chimpanzees, bonobos, gorillas, and orangutans living in zoos and sanctuaries, and compared them to similar measures from men and women from several different populations, including the U.S., South Africa, Jamaica, and the Hadza community. We learned a few things along the way, like the fact that some orangutans prefer to hold the urine-collection cups themselves, with their feet (and they often spill), and gorillas like their isotope doses mixed with sugar-free iced tea. We also learned that human bodies are evolved to get by with less water, about one-third less than our ape relatives.

We're still trying to understand all the ways our bodies conserve water. Our kidneys haven't changed much, as far as we can tell. One piece of the puzzle, we think, is that our big noses help retain water. Chimpanzees and other apes have flat faces, with hardly any nose. Humans, by comparison, have big honking schnozzes. When you breathe through your nose, much of the moisture in your exhaled breath condenses inside your nostrils, where it can be reabsorbed rather than lost. From fossil skulls, we can see big noses beginning to evolve around 2 million years ago, when early *Homo* species were adapting to dry savannas and learning to hunt and gather. The amount of water saved isn't huge, but evolution often works on small advantages and it could have been important, particularly in dry environments. In fact, we know that the pattern of nose shape across human groups today reflects variation in climate and humidity, something we talked about in chapter 2: populations with larger noses and smaller nostrils, which do a better job moisturizing the air you inhale and collecting water vapor you exhale, tend to live in drier climates.

Some of you are wondering how a card-carrying evolutionary anthropologist like me hasn't heard of the waterside hypothesis of human evolution. The general idea, sometimes called the aquatic ape theory, is that our early *Homo* ancestors didn't evolve in dry savannas but in and around lakes and rivers. This idea is incredibly popular in some corners of the internet, but hasn't caught on in mainstream human evolution research. There are many reasons for resistance to the idea, but I suspect the main reason is that it's wrong.

Aquatic ape enthusiasts often point to a long list of features throughout the human body that, in their view, are clear adaptations to life in the water. For every one of them, there is a simpler and more convincing case as an adaptation to land. Our legs and feet are evolved to walk and run, not wade and swim. There would be no selection to evolve water conservation if our ancestors lived in the water, and no advantage to sweating. True, fossils of our ancestors are often found in and around old lakes and streams, but that's the case for all species because that's where dead animals are mostly likely to be buried in silt and fossilized. Our Paleolithic forebearers thrived in nearly every habitat, and those that lived near water undoubtedly made use of its bounty. But there's no real evidence of an aquatic phase in human evolution.

The adaptations that allow us to survive in arid landscapes aren't just biological, they're also cultural. Our ancestors thrived in hot, dry climates by learning clever tricks to find water. I remember a hot afternoon in Hadzaland during one research trip there, walking with kids from camp to fetch water. The camp sat atop a hill, and I assumed as we left camp that we'd be making our way far down into the valley below—the only place I knew of in the area that had dependable water. Instead, we hiked a mile or so along the ridge to a huge baobab tree, and kids started climbing high into the crown, three stories up. I thought they were goofing off until they started lowering their buckets down into a hollow inside the tree's enormous trunk. Bucket after bucket was lifted back out with tea-colored water, which we carried back to camp for drinking and cooking.

Hadza, Daasanach, and other dry-climate populations also learn to pull water from the sand. Streams and rivers on these landscapes are completely dry for months on end, just ribbons of sand snaking through scrubby bushes and trees. With a sort of sixth sense that seems like magic to me, people learn to spot places where water running under the surface builds up. They dig wells by hand, five to ten feet deep, and inevitably hit water. They've even got effective tricks for cleaning out the silt.

That's not to say it's easy. Asher Rosinger, an anthropologist at Penn State and a colleague in the Daasanach research, has worked with the community to understand the impact and importance of water. It's a perennial source of worry and work. The burden falls most heavily on women, who do much of the physical work of fetching water and the social negotiation of borrowing from friends and family when buckets run dry. In these communities, fresh water can change lives. Studies tracking women in societies similar to the Daasanach have found that adding a well increases fertility. Water gathering is hard work, and such a drain on women's energy, that providing a well literally saves enough energy to build more babies. In our work with the Daasanach we're witnessing the impact firsthand and in real time, with a new desalinization plant providing clean water to parts of the region, while others continue to rely on hand-dug wells in the sand.

Asher's work has shown that individuals who are able to stay hydrated also stay healthier. Those health effects may even lead to local adaptation in water regulation. Genetics research in neighboring communities, by Amanda Lea and Julian Ayroles, suggests the consequences of dehydration have led to natural selection in kidney function. These hot, dry environments seem to have favored an allele of the *STC1* gene that enables the kidneys to produce more concentrated urine and retain more water.

In our modernized, industrial world, the clever tricks for finding water are all but forgotten, replaced by the engineering marvel of indoor plumbing. Yet somehow, we're still anxious about hydration. Some of this worry is warranted. Rosinger and his colleagues estimate that roughly 2 million people in the U.S. lack access to safe tap water, a burden that falls most heavily on poor and minority communities, as we saw in the nationally publicized debacle in Flint, Michigan. But the reality is that people in most places in the U.S. and other developed countries have clean, safe water right at their fingertips. Rosinger's analysis suggests 59 million avoid drinking water from their faucet for no clear reason, even though their water is safe, opting instead for bottled water or other drinks.

Societal anxieties about hydration are tied to ideas about purity and cleanliness. Many of us are convinced that we're being inundated with toxins. At the same time, we're bombarded with messaging that we're not drinking enough water, or that regular old tap water isn't enough to optimize our health, balance our pH, boost our metabolism, and generally make us immortal. There are legitimate reasons for concern (again, Flint, Michigan), but a lot of the worry and hype around hydration is ginned up by companies looking to sell you liquid magic. The bottled water business makes more than $300 *billion* per year globally. The detox drink market makes about $5 billion.

We'd be much more selective in our worry and resistant to this sort of mass-marketed-water hysteria if we understood how our kidneys work. They've been keeping us and our ancestors perfectly hydrated and toxin-free for millions of years.

KIDNEYS MIGHT BE THE MOST AMAZING, LEAST APPRECIATED ORGAN in your body. You've got two of them, each about the size of your palm, sitting low in the back of your abdominal cavity, tucked under your love handles. Together, they clean every drop of blood, all six liters of it, more than thirty times a day.

To understand the job they do, you've got to understand that your blood is not some homogenous fluid, like Kool-Aid or motor oil. Blood is a stew, with tons of little bits suspended in a watery base. Most of the bits are good, essential even, like sodium and potassium and glucose. Hidden among all that treasure is a hodgepodge of waste: cast-off bits of worn-out cells, old degraded hormones, toxins and pollutants, and other stuff you'd be better off without. Your kidneys have to sift through all of it and pull out the junk while leaving the good stuff alone. At the same time, it's their job to keep the stew's recipe *precisely* the same, minute by minute and day after day, balancing the water, salt, and acidity to keep the mixture *just* right. Just a hair out of range, and the chemistry that keeps you alive will fall apart.

All this work is done by an army of microscopic sieves called nephrons. Each kidney has about 1 million packed inside, and they are shaped like baritone saxophones, with valve-like openings all along the surface. A thin artery plunges into the "mouthpiece" (technical name: Bowman's capsule) of each nephron. Once inside, the artery bulges into a sphere, giving it the appearance of a tennis ball hidden in a tube sock, before squeezing back down and exiting back out. That sphere, called a glomerulus, is full of holes. Blood spews out, carrying with it all the bits in the stew—sodium, glucose, amino acids, broken-down hormones, toxins. Everything except for cells and the largest molecules like fats, which are too big to fit through the holes, streams out of the glomerulus and into the nephron.

Each drop of blood is filtered through your glomeruli every forty-five minutes or so. If your nephrons didn't put nearly *all* of it back into your bloodstream, you'd be in big trouble—your blood would be a thick jelly of cells and fat in less than an hour. This is where the "valves" of our microscopic nephron-saxophone come into play. The nephron is covered with pumps (the valves) that pull sodium, glucose, and other good stuff out of the stew. Water follows sodium along like a dog on a leash, a chemistry trick called osmosis. The long hairpin curve at the bottom of the nephron, called the loop of Henle, serves to squeeze more water and sodium out of the stew as it passes through. Water, sodium, and all the other good stuff gets pumped *back into* the arteries, after they've left Bowman's capsule, to go off to the rest of the body clean and fresh.

The little bit of fluid that's left in the nephron, along with toxins and other waste, exits out the end and into a network of collecting ducts that connect it to the bladder. There it will stay until your next trip to the bathroom, golden urine that was very recently the red blood flowing through your veins. The hormone vasopressin helps to control the volume of urine you produce. When the body senses it's getting dehydrated the hypothalamus produces vasopressin, which opens additional channels in the collecting ducts and pulls out more water, reducing the amount of urine that ends up in your bladder.* It's this mechanism that seems to be under selection in African pastoralists, helping people living in hot, arid climates to reduce the water they lose each day as urine.

The nephrons are exquisitely sensitive to the amount of oxygen and water in the blood, and they keep these critical variables in

---

* The body naturally produces vasopressin at night, which reduces the need to wake up and pee. Some kids who have trouble with bed-wetting don't produce enough vasopressin, and treatment with synthetic vasopressin can solve this issue.

balance by producing hormones that can ramp up blood volume or red blood cell production. Nephrons also help to keep electrolytes like sodium and potassium in balance. And they maintain the pH of your blood within a tightly controlled range, producing or excreting bicarbonate (yes, the same stuff you buy in a box of baking soda) as needed. If this all sounds like a lot of work, you're right. The cells in your kidneys burn more energy per day than the cells in any other organ.

In other words, the cleanses and alkaline water you're being sold are a waste of time and money. Your kidneys are hard at work taking care of everything already, just as they have been since you were born. And any toxins they can't get rid of, like the lead in Flint, Michigan's water supply, are well beyond the reach of homeopathic cures or magic detox rituals. The only thing cleanses

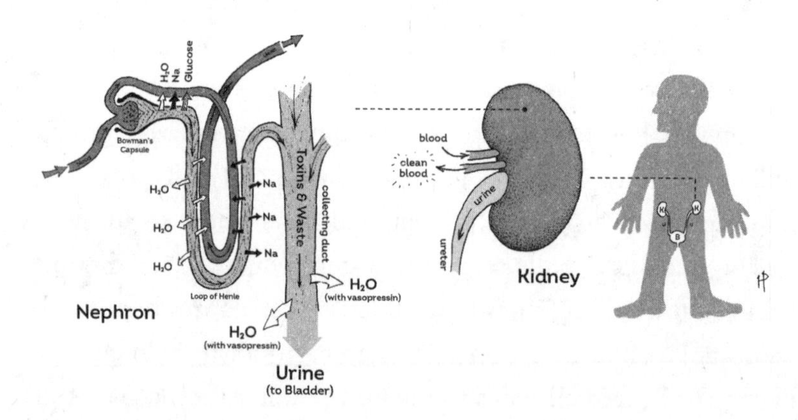

Figure 8.1 Every drop of blood is cleaned more than thirty times each day by millions of microscopic filters called nephrons, which are packed into each of your kidneys (K). Fluid from the blood streams into Bowman's capsule and into the nephron. Water (H_2O), sodium (Na), glucose, and other important molecules are pumped out of the nephron and collected back into the bloodstream. The mix of water, waste, and toxins that make it through the nephron is urine; it's funneled to collecting ducts, then to the ureter (u), and on to the bladder (B). More water can be pulled out of the urine (and the urine made more concentrated and darker yellow) with the hormone vasopressin, which allows water to escape from the collecting ducts.

remove is your money, and the only thing alkaline water balances is the supplement company's bank account.

As amazing as your kidneys are, you still need to drink. But how much? If you listen to the bottled water industry, the answer is always *more*! We're constantly reminded to "stay hydrated!" and offered an unending array of drinks and vessels to help us achieve liquid nirvana. Growing up, the benchmark I learned for proper water intake was eight glasses of water per day. And it had to be *water*—other drinks didn't count. I didn't know anyone who actually drank that much, and most people I know today still don't. Yet it's not uncommon these days to see people lugging around half-gallon carboys with motivational slogans to keep them guzzling.

You can't escape the messaging. I recently encountered a cheerful decal in a public restroom, stuck to the wall at eye level above a urinal, with a color-coded scale to assess your hydration. If your pee was anything short of crystal clear, it said, there was room for improvement. Alas, a quick check confirmed a disappointing yellow. I was water-shamed by a flowchart.

The good news, for those of us who cringe at the thought of drinking eight glasses of water or more each day, is that you don't need to do it. It turns out the eight-glasses target was largely pulled out of thin air by well-meaning folks in public health a few decades ago. It is more than a lot of people actually need, and you can get your water from nearly anything that has water in it. Coffee, tea, milk, juice, beer, and other beverages, not to mention nearly all foods, contain water that counts toward your daily intake. Only a few, like a strong shot of espresso or hard liquor, are likely to cause you to pee out more water than you consume.* The others are a net positive, adding to daily water intake.

---

* Alcohol and caffeine are "diuretics," meaning they reduce the production of vasopressin and cause your kidneys to produce more urine. Some medicines are

Water needs also vary from person to person. In 2022, my colleague Yosuke Yamada led a big international study to analyze water needs for thousands of adults around the globe, and found that most women consume about 100 ounces (2.9 liters) of water each day in their food and drink; men consume about 125 ounces (3.7 liters).* But the amount of variability was impressive, with some people consuming more than twice as much as others. Body size is a big factor (larger people need more), as is age (water needs decline along with your metabolic rate after age sixty). Climate and activity also influence your water needs. In hot climates, or if you're exercising, your water needs can easily increase by a liter or two. No surprise, then, that populations like the Hadza and Daasanach need more water each day than the typical American sitting at their desk in an air-conditioned office. In fact, Yamada's analysis found that people in low- and middle-income countries generally needed more water each day than those in rich countries, presumably because people there are more likely to work physically demanding jobs either outdoors or in buildings with poor air-conditioning.

So the amount of water you should be drinking will vary depending on your weight, how much you exercise, your age, and whether you're outside a lot on hot days. But if you're a typical adult in an industrialized country with climate control and relatively low levels of physical activity, it's highly unlikely you need to be drinking eight glasses of water each day. We can get up to 50 percent of our water through the food we eat, meaning that women might only need to drink about fifty ounces per day, men around sixty. With a couple cups of coffee in the morning, a

diuretics as well. Drinks with a large amount of diuretic in a small volume of liquid can cause a net *loss* of water from your body.

* Your body also makes water, about 300 ml (10 oz.) per day, as a by-product of making ATP in the mitochondria (see chapter 5). The oxygen you inhale becomes the O in H_2O.

zero-calorie soda for lunch, and an evening beer, the amount of water you need would be far less than eight glasses per day.

That's not an excuse to skimp. If you're thirsty, drink! Your internal hydration sensors do an excellent job of adjusting your level of thirst, first nudging you, then screaming at you, to drink. If you listen to your body, you should be in good shape. Chronically dehydrating yourself, even a bit, can cause trouble. Mild dehydration causes mental fatigue, headaches, and fogginess, and over the long term it's linked to an increased risk of kidney disease, urinary tract infections, kidney stones, and other problems.

But pushing yourself to drink more than you're thirsty for is unlikely to improve your health. In extreme cases, people who drink far, far beyond their thirst can consume water faster than their kidneys can clear it, lowering the salt concentration in their blood to dangerous levels. In one tragic instance, a woman running the 2002 Boston Marathon died from drinking too much water during the race in a misguided effort to stay hydrated. More often, drinking more water just means more trips to the bathroom. When people are asked to increase their water intake in intervention studies, they mostly just pee more. The water content of the blood doesn't change much; their kidneys just adjust to the increased water intake by producing more urine.

So drinking more doesn't (usually) hurt, but we can tone down the fearmongering around hydration. You've got an elegant and effective system evolved over hundreds of millions of years to keep your water needs perfectly in balance. That same system is doing a great job keeping your blood clean and toxin-free. If you're thirsty, *drink!*, and don't avoid tap water from the faucet if you don't really need to. Listen to your body, and you'll be fine.

WHITER SHADE OF PALE: SKIN COLOR AND THE INVENTION OF RACE

Dehydration and overheating aren't the only ways the sun can kill you. It has another weapon, invisible but effective: ultraviolet (UV) light.

The sun's energy hits the surface of the earth at midday with the intensity of a thousand-watt floodlight shining on every square meter (10 square feet). That energy comes in a wide range of frequencies, from X-rays to radio waves. The vertebrate eye evolved in the ocean, sensitive to the range of frequencies that could pass through the atmosphere and then through several feet of water. Our eyes, inherited from those ancient fishy ancestors, are still adapted to that same range of energies. We call the frequencies of energy our eyes can detect "light."

Dogs, cats, and most other mammals experience a sensory world dominated by smell and sound. Our primate ancestors, evolving in highly three-dimensional tree canopies to find and eat flowers and fruits, evolved to become visual creatures, depending more on what they could see. Real estate in the brain was taken from olfaction and ceded to processing visual stimuli. New alleles evolved for the genes that make the light-sensitive proteins (opsins) in the rod and cone cells lining the retina (the light-sensitive membrane in the back of the eye), enabling our monkey-like ancestors to distinguish different frequencies of light. We call those different frequencies "colors." The powerful visual processing cores in the back of your brain turn the scrambled chaos of color flooding your retinas every second into the spaces, objects, and beings that you call the world.

Curiously, there's a chunk of the visual world that we miss. As with other primates, the lenses in our eyes filter out UV light, frequencies that fall just beyond the blue-violet end of our rainbow. If you were to represent the range of light frequencies that most other animals can perceive as a set of twenty-six encyclopedia volumes, A through Z, the primate lens blocks the first eight, A through H. The loss seems to result from a trade-off with visual acuity. Many mammals, oriented to sound and smell and content with a low-resolution view of the world, can see UV light just fine, as can many (perhaps most) insects, birds, and fish. Primates, under selection to see a crisp, high-definition world, evolved lenses that filter UV out. We find similar UV-filtering lenses in sharp-eyed raptors that depend on their vision to hunt, catching prey on the wing.

Even though our lenses block UV light, the opsins in your retina remain sensitive to it, a holdover from our ancient ancestors. On rare occasions, some have been able to glimpse this hidden world of color. Impressionist painter Claude Monet had the lens in his right eye surgically removed in 1923; it had grown cloudy with cataracts, rendering him nearly blind. Without the lens he became sensitive to UV light, which he experienced as a whitish blue. Monet's expanded color palette found its way into his post-surgery paintings. Today, in modern cataract surgery, the cloudy lens is replaced with an artificial one tuned to the usual UV-filtering specifications. UV blocking was added to artificial lenses in the 1980s out of concern that UV light could damage the retina, although plenty of UV-sensitive animals don't seem to be bothered.

All the UV light we can't see absorbs into our skin, knocking molecules around like a rugby match in an art gallery. Some of this is good for you. Your skin is saturated with a molecule called 7-dehydrocholesterol, a by-product of your body's cholesterol management system, and UV light transforms that molecule into previtamin D. As the name suggests, previtamin D is the essential stuff for making vitamin D.

Vitamin D has a wide range of critical functions, including

building bones in growing kids and maintaining the balance of calcium levels in adults. Vitamin D's reach goes far beyond the skeleton, however, and there's some tantalizing evidence that healthy vitamin D levels might protect against depression and some forms of cancer. Clinical trials testing the impact of vitamin D supplements have been pretty disappointing for those outcomes, though, so the full impact remains uncertain.

You can get vitamin D naturally in some foods, particularly eggs and seafood. And one of the few nutritional bright spots in our strange, modern, processed-food diets is the addition of vitamin D into dairy products, from milk to yogurt. Prior to modern fortification, sunlight was necessary for making vitamin D in sufficient quantity.

But UV light is both Dr. Jekyll and Mr. Hyde, pleasant and helpful one moment and destructive the next. It tears apart molecules in your skin cells, including your DNA. Your cells have mechanisms to fix the damage, but every repair job carries with it the potential for error, a mutation in the DNA sequence that could be cancerous. UV light also shreds a molecule called folate, a B vitamin (there are several) that your cells need to make DNA. Folate is needed by all your cells, but it's particularly critical for a growing fetus and its trillions of cell divisions and DNA replications (making folate supplements a smart move during pregnancy). If the damage to your DNA, folate, and other molecules in your skin is severe enough, your body mounts a rescue mission, opening up blood vessels to the area and shipping in repair teams from the immune system. The inflamed, itchy, painful result is a sunburn.

Your body is caught in a delicate balance. You need enough UV to make vitamin D, but too much can cause skin cancer or harm a growing fetus. The nature of the challenge differs around the globe. At the equator and at high altitudes, where sunlight is most direct and the atmosphere is thin, you risk overexposure and damaged DNA. Toward the poles, where the sun hits the earth at an angle, you risk underexposure and insufficient vitamin D.

The evolutionary solution? Color, in the form of a molecule called melanin.

Melanin is ancient. It appeared so early in evolution that it's found throughout the tree of life today, from insects to flowers to us. It's a big, complex molecule with a number of useful chemical properties. Melanin binds to toxic metals, rendering them harmless. It's involved in the immune systems of reptiles and amphibians. But for mammals like us, melanin's most important property is also its most noticeable. It's a dark pigment that absorbs light, particularly in UV frequencies. By absorbing UV light, melanin prevents it from damaging your skin cells. Melanin is nature's sunblock.

As Nina Jablonski, an anthropologist at Penn State, details in her book *Living Color*, each of us produces melanin in our skin. Special cells called melanocytes produce little packets of the stuff, which migrate through the skin and park themselves near the nuclei of our cells, protecting the DNA inside from UV damage. There are two varieties: eumelanin, which is dark brown, and pheomelanin, which is more reddish yellow. Like Monet mixing colors on a palette, the proportions of eumelanin and pheomelanin determine your skin tone (coffee, peach, or salmon, for example), while the total amount produced determines how light or dark your skin is. Small concentrations of melanin color your freckles and moles.

Melanin is also notable when it's absent. People with albinism carry gene variants that interrupt melanin production, resulting in exceptionally light skin. Hair and eye color are often affected as well, since melanin is involved there, too. In people with vitiligo, the immune system attacks melanocytes in some areas of the skin, resulting in light patches.

Unlike most of the traits we've discussed in this book, skin color is strongly heritable. There are over 160 genes known to affect the production of melanin and therefore skin color. Some of those genes also determine whether your melanocytes respond to UV exposure by making more melanin, causing you to tan. Your particular skin color depends on the combination of alleles you have for each melanin-related gene and, if you've got alleles that enable you to tan, your recent sun exposure. Genes and environment, working together.

With melanin, evolution can calibrate the amount of UV light your skin absorbs. In places where sunlight is intense, natural selection favors alleles for darker skin, providing more natural sunblock to avoid damage. Where sunlight is weak, natural selection favors alleles for lighter skin, allowing more UV light in support of vitamin D production. In temperate latitudes, where sun intensity waxes and wanes with the seasons, natural selection favors the ability to tan in the summer and grow pale in winter.

The evolutionary pressure on skin color would have been particularly acute for our Paleolithic ancestors, as natural selection for sweating and thermoregulation robbed them of their body hair, exposing their skin directly to the equatorial sun. The optimal balance of light or dark would have varied around the globe. UV light exerts an evolutionary pressure that is strong, stable, and varies by location, the exact conditions that favor local adaptation. The earliest *Homo sapiens*, evolving in the tropics of Africa some three hundred thousand years ago, would have all had dark skin. But as our species grew and populations bubbled out of Africa and spread across the globe, skin color tracked the prevailing conditions. Populations developed lighter-skinned as they moved toward the poles, only to adopt darker again when subsequent generations moved back toward the equator. DNA evidence gleaned from archaeological populations suggests that the adoption of agriculture starting around twelve thousand years ago strengthened the selection pressures favoring lighter skin in low-UV regions, as people came to rely more heavily on farmed foods that were relatively low in vitamin D. Today, when we look at native populations around the world, the distribution of skin color is a faithful map of UV intensity. The alleles for melanin production are local, the classic example of natural selection leading to local adaptation in our species.

Take a moment to ponder the societal impact of this simple, elegant adaptation. Skin color and its ramifications—whether your ancestors were free or enslaved, whether your relatives today experience racism and its myriad impacts, your life expectancy and health—all result from a single, ancient molecule, melanin, and its role in protecting us from the sun.

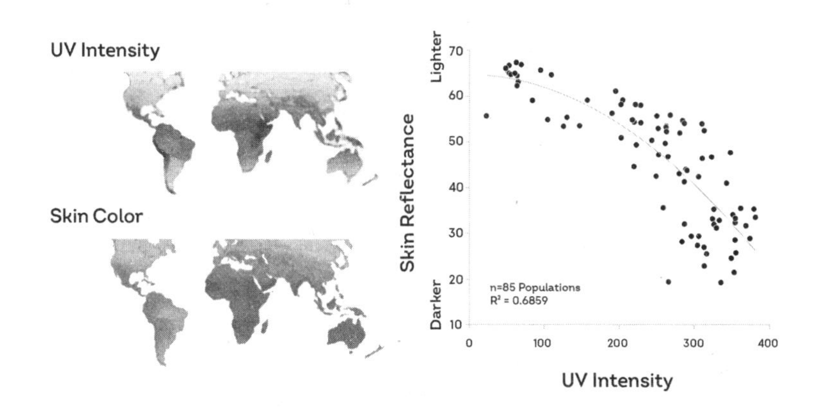

Figure 8.2 The intensity of ultraviolet (UV) light varies globally (darker areas receive more UV). Average skin color for Indigenous populations, such as the eighty-five populations from around the globe plotted in the figure on the right, corresponds to the global distribution of UV intensity.

But wait! We've spent most of this book talking about race as a cultural concept, unmoored from biology. Doesn't the biology of skin color prove the opposite? From the early classification schemes of Linnaeus and the horrors of slavery, to the modern discourse around "people of color" and Black Lives Matter, skin color has been integral to our concept of race. If skin color is biological, doesn't that mean race is, too?

The short answer is no, but to see why we need to understand a few essential points. First, skin color is what anthropologists and biologists call a "clinal" trait. Rather than hard-edged categories, there is continuous variation from lighter to darker as we move across the map, just like we'd expect from the continuous variation in sunlight intensity. When we see the full spectrum of color around the globe, it's clear that racial distinctions are culturally imposed, like splitting the world arbitrarily into "tall" and "short" people. Skin color doesn't divide the human species into categories, cultures do.

Another important point is that the genetics of skin color don't affect anything else of consequence. True, they influence eye and hair color somewhat, but even the genes that determine hair texture (whether your locks are straight, curly, or frizzy*) are independent of your skin color genes. The pattern of skin color variation doesn't tell us anything about the geographical distribution of other traits. For example, ABO blood groups vary around the world, but if we were to group regions by blood group we'd get a different set of blood group "races" than we see with skin color. Other traits and their alleles are the same way. You can have some fun visualizing this yourself with the University of Chicago's Geography of Genetic Variants Browser. Height, weight, enzymes, immune system alleles . . . if it varies globally, it's likely to give you a different pattern of variation. Yes, you can group people by skin

* Tightly curled hair, of classic Afro fame, is thought to have evolved as an adaptation to keep the head a bit cooler in hot climates, another great example of adaptation to local environments.

color (and skin color alleles) if you must, but you aren't capturing anything else about people's biology.

The independence of skin color from other traits leads to the last point we need to keep in mind when we think about race. Skin color is independent of other traits because human populations have been mixing, migrating, and marrying for our species' entire three-hundred-thousand-year history. We've been stirring the gene pool since the beginning, moving alleles around the globe. We used to wonder how extensive these movements were, but recent advances in the science of ancient DNA have removed all doubt. As David Reich, an ancient-DNA expert, describes in his book *Who We Are and How We Got Here*, the DNA from skeletons at archaeological digs around the globe tells a story of unending movement of individuals and groups. Europeans today have ancestry from Asia and the Middle East, who in turn have had major migrations from Africa, which has been its own swirling cauldron of migration. The beat goes on, in every corner of the world, for the full extent of our species' history. For anyone skeptical that their beloved ancestors mixed with other groups every chance they could, remember that we have documented evidence of humans sleeping with Neanderthals and other species. Humans aren't picky.

Colonialism and globalization have only accelerated this intermingling. One of the most important pieces of science to emerge from the rise of retail genetics companies like 23andMe is just how widespread intermixing and intermarriage have been here in the U.S. in spite of legal restrictions against interracial marriage prior to 1967. Most Americans who identify as Black have some European ancestry, and a substantial number (more than 10 percent in parts of the South) of self-identified White people have recent African ancestry. It's likely that few of these folks know their mixed-race background, since such couplings were illegal and often hidden until recently. There's plenty of evidence of self-identified Black and White U.S. adults with Native American ancestry as well. Today, about 15 percent of all newborns in the U.S. are mixed-race, and that number is steadily growing. There are no pure groups, racial or otherwise, and there never were.

As a result of all this mixing, people and populations around the globe are remarkably similar genetically. You share more than 99.9 percent of your DNA with every other person on the planet. Pick any population in the world, say some group from the remote rainforests of central Africa, and over 90 percent of the alleles (gene variants) in that population will be found in every other population on the planet, across all continents and habitats. Not only that, but that remote African population is likely to be more genetically similar to a population in Europe, Asia, or elsewhere than it is to another randomly chosen population in Africa. If we were to use genetics to divide our species into groups, we'd find very little to base those distinctions on, lots of overlap between groups, and a set of categories that didn't look like the racial groups we're familiar with.

And the handful of alleles that aren't shared among populations? Only a tiny fraction of that variation has any meaningful impact on biology, like the variation in body proportions that we find with the Daasanach or adaptations to malaria resistance in some equatorial populations. Importantly, those variants aren't shared across an entire "race" (for example, across all of Africa, or all of Asia), but are much more localized and calibrated to particular environmental pressures.

Most human variation is neutral. It's meaningless noise, the result of random genetic mutations, the mixing caused by migration and intermarriage (what geneticists call "gene flow"), and the random loss of variants in the lottery of passing down alleles (what geneticists call "genetic drift"). The accuracy of ancestry testing and the distribution of variants around the globe might seem to support the idea that groups differ genetically in important ways, but that gets the science exactly backward. Those variants are helpful for tracing ancestry precisely because they're biologically meaningless. Random, neutral variants that arise in one group are unlikely to occur in another, providing a fingerprint-like uniqueness to the genome. Variants that affect survival and reproduction can be favored (or disfavored) in many populations, and therefore aren't nearly as helpful for tracing ancestry.

Understanding the biology of skin color and the cultural invention of race should force a reckoning in the ways we think about our bodies and our health. Race is cultural, but we make it biological by treating people differently depending on their race. Biologist Joseph Graves and biological anthropologist Alan Goodman detail this phenomenon in their book *Racism, Not Race.* In the U.S., the long history of racial oppression and policies favoring White people has led to systemic differences in the social and physical environments in which people live. Black, Hispanic, Indigenous, and other communities of color have less wealth, less access to education, less access to health care, and other well-documented disadvantages. They also face racism in their social and professional lives. These cultural and economic disparities lead to racial disparities in our biology, with negative consequences for health across nearly every metric, from obesity to heart disease to life expectancy.

Racial health disparities arise from racism, not genetics or other innate differences. Yet race is frequently used in medicine and medical research as though it were an innate, biological variable. Equations that doctors use to calculate your kidney function score, for example, include a variable for whether you are Black. Why should the melanin content of your skin or the racial identity you ticked on your intake form have anything at all to do with how well your nephrons are working? The obvious and correct answer is *they don't*. But algorithms like these, ubiquitous in medicine, perpetuate the mistaken idea that race reflects innate biological difference. If there are systemic racial differences in kidney function or any other trait—and there often are in places with a history of racism—we should look to environmental causes rather than assuming they're genetic.

The rest of us, outside of the medical world, must continue to confront our misguided folk wisdom around diversity, the ugly side of our cultural inheritance. We evolved as the sharing ape, and can be incredibly generous with others in our community. But the flip side is a disregard and even hostility toward people who are not in our community. Who belongs in our community and who does not is something we absorb as part of the Story of How the World Works, passed down from generation to generation. Too many of us are taught from an early age to notice superficial variation and read into an unbridgeable chasm of difference. Much of this learning is passive, our developing minds subconsciously building categories to make sense of the world. As primates, we lean heavily on differences we can see because of our ancient, evolved dependency on vision: skin, noses, eyes, hair, clothes . . . any cue is fair game. Babies learn to use skin color to identify faces before their first birthday, and children are aware of racial differences in appearance before they're in kindergarten.

When differences don't exist, we invent them. As Jablonski details elegantly in *Living Color*, civilizations the world over have been dividing the social world into classes, castes, races, and other groups for millennia, often (but not always) using skin color. Skin color was of course central to the construction of race in the Americas, with the distinction between White and Black codified in the late 1600s. White European settlers used skin color to dehumanize and justify the enslavement and cruel treatment of Black Africans. But for anyone of mixed race, light skin was not enough to be considered White. Those in power went to great lengths to categorize an increasingly diverse American population and exclude people from the social and economic benefits of being White. Anthropologists in the late 1800s and early 1900s quantified meaningless variation in heads and bodies and sorted people into categories in the service of the racist and eugenic efforts that carried the day. Italians, Irish, and other light-skinned Europeans weren't viewed as White by the entrenched political class until the mid-1900s.

The cultural construction of race continues today, in the U.S. and abroad. The evidence isn't difficult to find. Cartoonists and propagandists still use tropes about noses, hair, and other traits to invoke differences between groups. And just as in the past, these caricatured differences, often invented out of whole cloth, are used to dehumanize and oppress. In the Rwandan genocide of 1994, the Tutsi viewed their Hutu neighbors as darker and shorter, helping to make them different enough to justify their execution. Sectarian and religious differences in the Near East are central to the seemingly endless conflicts we see there today. We have a tendency to divide our social universe into Us and Them, and we will find or invent differences to justify our distinctions. It's a tendency we must learn to resist.

Our incredible success as a species has come in large part from our abilities to engineer our environments. We figured out how to manage fire, bringing light and warmth into the cold and dark. We changed dozens of plant and animal species through selective breeding to provide food and labor. We built homes, then camps, then towns, then cities, transforming wild ecosystems into landscapes of glass and steel. Today, we have the ability to transform the planet at will.

We are just waking up to the unintended consequences of our environmental engineering. Humans are warming the oceans and the atmosphere to levels unseen in thousands of years, with a trajectory that seems likely to melt the poles, cause mass extinctions, and drown our coasts. Massive, modern metropolises have arisen in hot, dry regions that are only going to get hotter and drier. We are pushing our bodies and our ability to manage heat and hydration into uncomfortable and dangerous territory. It remains to be seen whether we've come to understand the gravity and scope of the climate crisis in time to prevent mass misery and societal dislocation.

We've engineered our social environments as well, intentionally and unintentionally, in ways that have systemically hurt communities of color and other minority groups. It is no small irony that melanin, something we all share and that adapted to protect us from threats in the natural environment, became a weapon of division and oppression in our human-made cultural environments. Undoing all the harm caused by racism in the U.S. and elsewhere has proven to be a multigenerational challenge, and the work is far from over. But like climate change, the problem is too big to ignore and too important to concede. We need to continue and support efforts to end racism and repair the damage caused. And, like the climate crisis, we'll need long-term commitment. Environments, social and physical, take time to change, and as we've seen throughout this book, early environments can have effects that last a lifetime and even spill over into the next generation.

There's reason for hope. A primate clever enough to reshape a planet certainly has the power to engineer its social and physical environments to be fair and sustainable. It would be nice to live long enough to see that through. Is there a chance we might? While we're engaged in all this world making, can we engineer away all the other bad stuff, too, like disease and frailty? Can we find dependable ways to repair and rejuvenate our bodies when they break down? In short, can you live forever?

9

HOW NOT TO DIE (AND WHY YOU WILL ANYWAY)

There's a school of thought, as old as Rousseau and still popular on social media, that we'd all be happier and healthier if we returned to a hunter-gatherer way of life. It's the Paleo diet writ large. Groups like the Hadza are the North Star for these Paleo fantasies, romping about in the soft focus of some imagined Eden, free from processed foods, alarm clocks, and modern medicine. I'm sympathetic, to a point. Eating less processed food and getting outside and active every day is a great idea, and the trappings of industrialized life make that hard. But there's a lot to like about the modern world, things we take for granted. Hunter-gatherer life is no picnic. Sometimes you end up with an eyeball in your leg.

A Hadza man in his early twenties, let's call him H, had just returned from a day spent tracking a wounded giraffe. Brian, my buddy and the leader of our research team, had spent the day with him, and had come back with concerning news. H had been complaining of a headache and tiredness, saying he'd felt off for a couple days now. He figured it was related to a cut he'd gotten in his thigh. It was deep and wasn't healing well, and the flesh around it had grown tender and swollen, in step with his headache. Brian asked if I'd have a look, as a former EMT and the research group's unofficial first aid guy.

Even with my limited training I could tell it wasn't great. There was clearly a knot of infection growing in H's thigh, an abscess under the skin. It seemed possible the infection was causing the lethargy and headache. Left alone, he might pull through . . . or he might go septic and die. Like young men everywhere, H had no interest in seeing an actual doctor, which was far away. If we were going to treat it, we were on our own.

It seemed a good idea to disinfect the wound—we had iodine and antibiotic ointment in the first aid kit. First, though, we needed to drain the abscess. H agreed, and after some discussion and cleaning the skin, he went to work. He pressed hard around the edges of the wound with his thumbs, squeezing out the infected ooze. After a few painful minutes, white cakey gunk started to come out in pieces, like feta cheese crumbles. H grew agitated and stopped.

"What about the eyeball?" he asked.

"Huh?" I had no idea what he was asking. Was he delirious?

"The eyeball. There's an eyeball in there, and if we take it out, there's going to be a hole. . . ."

I started to understand. He thought the white lump of infection in his thigh was an eyeball, or at least eyeball-like. A working understanding of the body, or even just eyeballs, would have been helpful, but this was no time for a lecture. Removing an eyeball is not to be undertaken lightly (everyone knows that) and we needed a good reason to deviate from the standard protocol we all learn from a young age, which is to leave eyeballs alone.

"We'll replace it with medicine," I said, trying to find an

explanation that was both truthful and convincing. "You pop out the eyeball, then we'll put medicine in the hole."

H pondered that for a moment, then went back to work. Eventually, infected bits stopped coming out. I mixed up a tincture of iodine and clean water and flushed out the wound, then packed the strongest antibiotic cream we had into the hole (you know, to replace the eyeball), and we wrapped him up. In a day or two he was back to his old self, tromping about the savanna.

It doesn't always work out that way. Infection is the number one killer in farming and foraging communities without access to modern medicine. Nearly half of all babies born in these populations die before the age of twenty. Adults succumb as well, especially the elderly, dying from pneumonias and other infections that we routinely prevent or cure in the developed world. But vaccines and antibiotics, triumphs of modern medicine, have become victims of their own success. We've gotten so good at treating infectious disease that we've forgotten how bad they can be.

COVID was a wake-up call to just how quickly infectious disease can bring us low. Yet, once again, the rapid development of effective vaccines and other treatments have led to a sort of cultural amnesia. Like H, most people have a spotty understanding of how our immune systems work. That collective ignorance has been fertile ground for conspiracy theories, growing online and in our politics like infected eyeball-lumps. A working understanding of our bodies, and in particular our immune systems, is a powerful remedy.

YOU AND WHAT ARMY?

Your immune system resembles nothing so much as a military, with billions of cells tasked with monitoring and destroying anything that invades. Most of these cells are white, giving the mass in H's leg its eyeball-like appearance. That's why the immune system cells in your blood are often referred to as white blood cells (or "leukocytes"). Like a military, different groups of immune cells have their own specialties, with a first line of defense backed up by seasoned veterans. Your immune system has a memory, a sort of military intelligence, readying it to respond to repeat attacks. It even has a sort of counterintelligence agency to root out healthy cells that go rogue and cancerous.

All your immune cells (and there are roughly a dozen types) are born in your bone marrow, from the same cell groups that pump out red blood cells. They leave home to lurk in your bloodstream or organs, living among your other cells like bounty hunters. Immune cells are blind, like all cells, sensing the environment around them through feel. Healthy cells from your own tissues have a protein structure on their surface called an MHC complex, a sort of ID card. Foreign cells, including bacteria, don't have your personal MHC complex and are attacked as intruders.* Infected cells often place bits of virus or bacteria on their MHC complex, notifying immune cells of trouble. They're calling for their own destruction. Many bounty hunter immune cells are "phagocytic," meaning they engulf and digest intruders and infected cells. Others are "cytotoxic," releasing molecules that kill any offending cell they encounter.

MHC identification also helps your immune system fend off cancer. The DNA mutations that cause a cell to become cancerous, dividing heedlessly and forming tumors or gumming up the blood with a torrent of excess cells, often corrupt the MHC complex. Those MHC changes flag them for destruction by natural

* People who receive transplanted organs or tissues need to find donors with MHC complexes similar to theirs, and even then they typically take medications to suppress their immune system and prevent their immune cells from attacking cells in the transplant.

killer cells and cytotoxic T cells, immune cells that specialize in taking out cancers.

When immune system cells sense an intruder they sound the alarm, secreting a wide range of molecules that stimulate other parts of the immune system military. Some of these signals are sensed by the hypothalamus in the brain, causing it to increase the body's temperature setting and induce a fever to help kill off invaders. Many of the signals cause inflammation, opening the capillaries to flood the infected region with more immune cells. Inflammation is what caused H's wound to be swollen and tender, and the millions of immune system cells rushing to the injury formed the bulk of the abscess. The wave of fluid and immune cells helps knock out an infection quickly, before it causes too much damage. Sometimes, as with H and his abscess, it develops into a prolonged battle.

The mismatch between the environments in which we evolved and the world we live in today seems to wreak havoc with our evolved inflammatory response. Our modern diets and sedentary lifestyles promote chronic inflammation that has no obvious target and never shuts off, damaging our blood vessels and other tissues. Obesity is a major factor, as excess fat promotes chronic inflammation. There's some evidence that ultra-processed foods full of added sugars and fats can increase the production of inflammatory molecules like C-reactive protein (often called CRP) as well, and as we discussed in chapter 6, they also seem to promote obesity. Exercise—really, any physical activity that gets your heart rate up—can reduce inflammation, but we in the industrialized world don't tend to get enough. Chronic inflammation is a major contributor to heart disease and other leading causes of death and misery in the developed world. In a bit of cosmic irony, our immune systems, evolved to keep us alive, are killing us.

Modern mismanagement of our immune systems goes beyond

the usual suspects of diet and exercise. It seems we've also created trouble by wiping out the germs that used be a part of daily life. Many of the bacteria and other pathogens we encounter are harmful, but others become part of our body's ecosystem, our microbiome. Microbiome bacteria are found all over and inside your body, with the largest population of them in your large intestine, where they dine on fiber and other bits of your diet that your own enzymes can't digest. In our well-intended push to reduce infectious disease, we've vastly reduced the number of germs we encounter, both friend and foe. Our immune systems, which are constantly reacting and adapting, calibrating their responses, develop in a strange, unstimulating environment. Like an enormous military that was built to fight world wars but now faces no credible threats, the immune system ramps up its presence everywhere and overreacts to minor infractions. Collateral damage and friendly fire are all too common. It's an idea called the hygiene hypothesis, first articulated in the 1980s and refined and revised since.

Allergies like hay fever are the poster children for the hygiene hypothesis. Histamine, a powerful inflammatory molecule produced by your immune cells, is a normal part of a healthy response to intruders. But in many of us, our systems overreact to dust, pollen, or other harmless stuff and produce massive amounts of histamine unnecessarily, causing the familiar runny noses, scratchy throats, and sinus headaches of allergy season. In asthma, this reaction constricts the airways in the lungs, making it hard to breathe. These and other histamine-driven reactions to airborne allergens were essentially unknown before 1900 and have increased five- to tenfold since the mid-1900s. Food allergies have taken off more recently, as anyone who has navigated the minefield of packing a schoolkid's lunch knows all too well. These allergic (over)reactions can be fatal, the swelling closing off your throat and airways and suffocating you.

Immune misfiring is also evident in the many autoimmune diseases that appear to be on the rise, from Crohn's disease to rheumatoid arthritis. In these diseases, immune cells attack innocent cells from our own body, damaging the tissues and causing trouble ranging from mild discomfort to debilitating disease. The triggers that set autoimmune diseases into motion aren't well understood, but the proportion of people with autoimmune disease has been growing in the developed world for decades, and is associated with a decrease in exposure to germs. One recent U.S. study found the number of people with antinuclear bodies in their blood, a reliable indicator of autoimmune disease, had more than doubled since the 1980s, while the prevalence of common infections declined.

The solution to the rising tide of allergies and autoimmune diseases isn't clear. Hygiene, from soap to clean water to sewer systems, has been an unparalleled success in public health. In the early 1800s, forty-five out of every one hundred kids born in the U.S. died before the age of five. Today, because of antibiotics, vaccines, and cleaner environments, that number is less than one. We don't want to go back to the old days. But it might be a good idea to get kids outdoors, exposed to more of the world's germs, pollen, and dirt. Send them to day care and elementary school so their immune systems encounter germs early, and vaccinate them so they can train their immune systems without getting dangerously sick. But keep up the handwashing. Let your kids get muddy and play with the neighbor's dog, then get them cleaned up before dinner.

LEARNING TO DEFEND YOURSELF

Microbiomes are just one facet of our complicated relationship with other species. We love animals and we eat them; breed them and push them to extinction; conserve some species in their natural state while reshaping others completely. When we try to apply what we learn from other species to ourselves, the results are equally mixed. Darwin looked at the natural world and saw in the species around us our deep connection to all living things. Then people like Francis Galton and Lewis Terman combined the new science of evolution with the old practices of animal husbandry to create the modern horror of eugenics. We've made major scientific discoveries from lab studies of mice and rats, but for every breakthrough there are plenty of promising results that simply don't translate well to humans.

Our close relationship with other species has also introduced new ways to die. Smallpox, rabies, HIV, the bubonic plague, malaria, COVID . . . the list of world-altering diseases that jumped to humans from other species, or learned to hide in the animals around us, is long, brutal, and ever growing. As I write this, a highly contagious form of bird flu has jumped to cows, and there's concern of it evolving to infect humans. On the other hand, our communal animals and the diseases they've brought us have been instrumental in the development of one of our greatest medical breakthroughs: vaccines.

Louis Pasteur, of "pasteurized milk" fame, built the foundation of modern vaccination with his work on rabies. Rabies is a devastating virus that's transmitted in the saliva of an infected animal's bite. From the wound, it creeps slowly up the nerves and into the brain, causing it to swell and turning the victim delirious and violent before killing them. In the mid-1800s, rabies was having a real cultural moment in France (yes, reader, it had gone viral). People were understandably terrified of the disease, which is one of the deadliest on the planet, killing more than 99 percent of its victims. Pasteur, already famous for his work establishing that germs (not spirits or smells) cause disease, decided to dedicate the final chapter of his career to developing a rabies vaccine.

The idea of vaccination was already centuries old by the time Pasteur took this project on. People in Africa and Asia had figured out that early exposure to smallpox, for example, rendered people immune from the disease later in life. These early cultures established the practice of inoculation, a crude but effective kind of vaccination in which a scalpel dipped in pus from an infected person's lesion was dabbed into a small incision on an uninfected person. The idea was imported to Europe in the 1700s, largely due to the advocacy of Lady Mary Wortley Montagu, an English aristocrat who had suffered smallpox as a child and learned of inoculation during her travels to Turkey. By the mid-1700s, inoculation for smallpox (also called "variolation") was widespread throughout Europe and the American colonies. George Washington made it mandatory for his troops. Leaders in Boston pushed inoculation despite protests from an uneducated and skeptical public. When a wave of smallpox swept through the city in 1721, the mortality rate for those inoculated was 2 percent, compared to 15 percent for those who weren't.

Pasteur had experimented with inoculation in the 1870s, developing vaccinations for chicken cholera and anthrax in farm animals from less potent (but still living) cultures of the diseases grown in the lab. He tried a similar approach with rabies, but came to realize he had killed his lab-grown stock of rabies viruses prior to injecting them as a vaccine. Rather than simply being less infectious, the viruses in his vaccine were "inactivated": inert and unable to infect the recipient. (You could also say they were "dead," although that presupposes viruses are alive, which is a matter of debate in biology.*) His rabies vaccine worked, saving the life of a nine-year-old boy in 1885 in a highly publicized early

---

* Unlike bacteria, which are self-contained living cells with all the parts needed to metabolize nutrients and reproduce, viruses are just DNA or RNA wrapped in a protein coat (sometimes with other molecules to promote infection).

demonstration and placing Pasteur firmly in the pantheon of scientific greats. He had invented the first modern vaccine, paving the way for effective and safe vaccines that would change the planet, virtually eliminating diseases, from smallpox to polio.

It would take another eighty years to figure out *how* vaccination works. Vaccination activates the body's "acquired" or "adaptive" immune system, which might just be the best buddy cop movie you've never seen. Two classes of immune cells, B cells and helper T cells, are longtime partners, monitoring the streets for trouble. Helper T has a network of informants, the phagocytic immune cells out there on the front lines battling bacteria and other intruders, collectively known as "antigens." Those informants present bits of antigens to helper T, who then shares them with his partner, B cell. B jumps into action, dividing into two kinds of cells: plasma and memory cells. Plasma cells produce trillions of proteins called "antibodies," each tailor-made to fit the specific antigen intruder. Antibodies flood the body, latching onto any infected cell with that particular antigen stuck to it and marking it for destruction. Memory cells hang out, lurking in the background, poised to jump into action and produce more plasma cells and antibodies if that antigen is ever encountered again. Together, the helper T and B cells fight off infection today and prepare the immune system to react even more quickly the next time.

Early inoculation approaches, like those for smallpox in the 1700s, prime the helper T and B cells by infecting the body with a small amount of virus or bacteria, or with a closely related, less virulent strain. The recipient would have a minor illness but be protected from full-blown disease later on. The genius of modern vaccines, like Pasteur's for rabies, is that they provide a molecular decoy, an inert antigen that has the shape of a nasty virus or bacteria, but is actually harmless. The immune system response, with the helper T and B cells working together to make antibodies, might activate other immune responses like inflammation and fever, causing you to feel crummy. But you won't get the actual disease, and if you are ever exposed to the real enemy you'll be primed to overwhelm it quickly with a massive antibody response.

With mRNA vaccines, like those developed for COVID, the shot doesn't inject you with antigens (inactivated viruses). Instead, the shot contains instructions to *make* specific parts of the antigens. It's such an incredibly clever bit of engineering, its discoverers won the 2023 Nobel Prize in medicine. If you're a bit hazy on your cellular biology, mRNA is the messenger molecule that travels between the DNA in the nucleus of your cells and the cellular machinery that makes proteins (see chapter 2). It's like a template, transferring the exact sequence of a gene in the DNA to the protein-building parts of the cell. The new class of mRNA vaccines inject mRNA sequences, built in the lab, which your cells then use to make proteins that mimic parts of a virus or bacteria. With COVID mRNA vaccines, for example, the injected mRNA template produces the distinctive spike protein that covers the outside of the COVID virus. These antigens are picked up by helper T cells, used by B cells to make antibodies, and voilà: you're prepared for a COVID infection without ever contracting the virus.

Vaccines have proven incredibly effective for training the immune system to fight a wide range of diseases. We've also figured out how to use them to target cancer cells, ushering in a new era of effective treatment for diseases that had no cure. You might think that vaccine researchers and the public health workers that distribute them would be hailed as heroes, and they are, usually. Sadly, through misinformation and fearmongering, these same efforts have also been derailed in some cases, with vaccine proponents targeted as villains.

Vaccine skepticism is as old as the technique itself. We can't see viruses or bacteria, and we generally do a bad job of estimating our risk of future disease. What our primate brain reacts to most potently is the here and now, the pain of an injection or the sore arm and crummy feeling that can follow a jab. We're also prone to see patterns in random associations, and quick to assign blame.

In 1721, anti-vax terrorists, driven by religious extremism and ignorance, firebombed the home of a prominent vaccine advocate who was leading one of the earliest vaccination campaigns against smallpox. Anti-vax efforts seem to have quieted in the early and mid-1900s, as modern vaccinations saved millions from smallpox, tetanus, polio, measles, rubella, and other childhood killers. But misgivings and misunderstanding never fully went away. When charlatans began scaring people that childhood vaccines supposedly cause autism in the 1990s, many anxious parents were quick to believe them. Mothers and fathers of children with autism wanted an explanation for their child's problems and were drawn to the coincidental timing: autism symptoms are often first seen around age two, the same time that kids are receiving standard, lifesaving vaccines. The autism-vaccination propaganda was built on lies, and the connection between vaccines and autism has been thoroughly investigated and completely debunked, but the myth lives on.

Vaccine hesitancy raised its head again with the COVID pandemic. People don't understand how vaccines work, and in the garden of ignorance a colorful riot of conspiracy theories has bloomed. For the record, mRNA vaccines don't make you magnetic, or give you COVID, or allow Bill Gates to track you from satellites (and why would he bother, when you're completely trackable from your smartphone?).

Ignorance around vaccines and immune systems is compounded by our general inability to assess risks and probabilities, something we discussed with heart disease in chapter 4. We assume we won't get sick, or that it won't be too bad. We weigh the hypothetical chance of being ill (or worse) against the inconvenience and discomfort of a shot, and we make the wrong call. Like the philosopher's trolley problem, where we're asked if we'd throw a switch to direct an out-of-control railcar toward a single person if it would save dozens on the car's current track, we hesitate. Taking an action feels different than standing aside, even though both are conscious decisions that carry their own set of risks. We avoid taking the positive step of getting a vaccine without fully realizing that *not* getting vaccinated is also a decision, one with a much steeper risk profile.

Fearmongering about vaccine side effects doesn't help. It's true that there is some vanishingly small risk of a bad reaction to the jab, just as there is with taking aspirin or other common and safe medications, but these risks are inevitably blown out of proportion. The scary headlines and social media posts about negative effects of vaccination hide the fact that people also respond poorly to placebo injections. If you do *anything* to a few million people, whether it's a lifesaving injection or a pat on the back, it is inevitable that some small number will have something bad happen to them in the following hours or days. A recent review examined over forty randomized control trials, the most scientifically rigorous approach to testing effectiveness and safety, comparing mRNA and other COVID vaccines to placebo injections. They found that the rate of "serious adverse events," like heart inflammation (myocarditis) or other major complications, was incredibly low (less than 1 percent) and didn't differ between placebo and vaccine, regardless of vaccine type. The effectiveness of mRNA vaccines was undeniable, reducing the risk of severe COVID by over 90 percent. In the U.S., the probability of being hospitalized and dying from a COVID infection is more than ten times greater for unvaccinated people.

Even for children, who have a much lower risk of serious complications or death from COVID, the benefits of vaccination outweigh the risks. A 2023 study in the *Lancet* examined COVID vaccine effectiveness and safety in children five to eleven years old, reviewing data from fifty-one studies and millions of kids. In their analysis, which included large-scale vaccine programs as well as randomized control trials, the rate of serious adverse events was indistinguishable between vaccinated and unvaccinated children. And while a jab and a booster didn't offer perfect protection against getting COVID, they did reduce the risk of severe symptoms and hospitalization by more than 50 percent.

Scare campaigns and conspiracy theories around vaccines, from the 1990s through today, have had deadly consequences. Childhood vaccination rates are falling in countries around the world, leading to outbreaks of diseases that had been effectively eliminated. Outbreaks of measles occurred in the U.S. in 2018 for the first time in decades. Worldwide, two hundred thousand people died from measles in 2019, up 50 percent from numbers in 2016. It's even affecting how we care for our pets. Vaccine hesitancy is growing among U.S. dog owners, leading them to opt out of legally mandated (and completely safe and effective) rabies vaccines. Pasteur, who became inactivated in 1895, must be rolling in his grave.

As with everything these days, there's a strong element of political polarization in these outcomes. Vaccine hesitancy has become a core issue for many on the political right. In the U.S., people living in right-leaning Republican counties are less likely to get vaccinated against COVID, and more likely to die from it, than people in counties that vote Democratic. A recent analysis in the *Journal of the American Medical Association* shows how anti-vax politics play out at the individual level. Before the COVID epidemic, registered Democratic and Republican voters in Florida and Ohio had equivalent mortality rates, and the numbers remained similar through the early months of the crisis. But once vaccines were available, their fortunes diverged. Republicans, no doubt influenced by anti-vax leaders on the political right, began dying in greater numbers. By 2023, excess deaths were 10 percent higher among Republicans in Ohio and Florida compared to their Democratic peers.

A common anti-vax rallying cry since the 1700s is that we should let nature take its course. That approach *might* eventually work, but the only guarantee is a lot of unnecessary suffering. It requires a lot of people to get sick, for one thing, putting them at risk of long-lasting problems or death, and the antibody response to real disease isn't any better than it is to an effective vaccine. And it's far from certain that rampant infection will produce society-wide protection. Smallpox, to give just one example, haunted our species for ten thousand years until we finally put an end to it with vaccines.

When we do see populations adapt to disease, it comes at an incredible cost. Diseases can exert powerful evolutionary pressures, often leading to genetic adaptation, as we saw with antimalarial sickle cell alleles for red blood cells in chapter 4. But every immune system adaptation in our DNA today is a memorial to the countless dead and disabled who didn't carry the lucky allele in the past. When the bubonic plague swept through Europe in the 1300s, it killed between 30 and 50 percent of the population, a horrific event that came to be known as the Black Death. From DNA analyses carried out recently in cemeteries from that period, we now know that variants of at least four different genes, active in immune cells, influenced a person's likelihood of survival. Today, those lucky alleles are far more common in European populations than they were before the Black Death. The modern population might be better protected from the plague, but only because so many people without those alleles died. I'm guessing they would have preferred a vaccine.

WHY NOT LIVE FOREVER?

Immunity is a moving target. Your personal protection against disease combines the genetics you inherited with the experiences you've had encountering bacteria, viruses, and other antigens, including those introduced through vaccines. The alleles you carry at hundreds, if not thousands, of genes will influence your ability to fight off disease, and the germs you meet each day train your immune system, including your learned antibody response. Lifestyle affects your readiness to fight as well. Immune response requires energy. Good nutrition and regular exercise help to keep your body prepared, while chronic stress and lack of sleep can dampen your immune response.

It's tempting to try and hack your health by adding the latest superfood or supplement, but there's no evidence they actually help. Unless you're eating a highly specialized and restricted diet or have some other known deficiency, you're unlikely to see much of an immune system boost from multivitamins, for example. Even the venerable vitamin C doesn't seem to offer additional protection against the common cold for the large majority of us who are already meeting our basic nutritional requirements.

Nonetheless, by some measures, immortality seems tantalizingly within sight, even if it's not yet within our grasp. We've added decades to our life expectancy in the past century, and the list of diseases conquered by modern medicine grows every year. It's true that modern plagues have arisen, like heart disease and diabetes, but even these seem to be slowly, finally, responding to modern treatment and prevention. In a population like the Hadza or the Daasanach, we often find elders in their seventies and even a few in their eighties, but I've never met nor heard of anyone in those societies living into their nineties. In the developed world today there are nearly six hundred thousand people over one hundred years old, ten times more than just fifty years ago. Could we keep pushing the limits to two hundred years, five hundred years, or more? Is it possible that one day aging will be just another disease that we can treat and reverse?

Before you shake your head, consider this: nature has already figured out how to do it. As Steven Austad, an expert on the subject of longevity, writes in *Methuselah's Zoo*, lots of species have evolved lifespans far longer than ours. Bowhead whales, Greenland sharks, sponges, and other animals regularly live for multiple centuries. Several tree species are known to live for *thousands* of years. There's even a species of jellyfish, *Turritopsis dohrnii*, that never dies. When times get tough, it ages backward, Benjamin Button style, taking on its juvenile form before throwing the transmission back into drive and maturing again. It can cycle like this indefinitely. If evolution has figured out immortality, using the same building blocks common to all life, perhaps it's not far-fetched to imagine that we might learn how to live forever, too.

Trying to live forever is actually two challenges. First, you need to run the gauntlet of disease that awaits us all: heart, lung, and vascular diseases, diabetes, cancer, infection, Alzheimer's disease, autoimmune disease, and others lurking in the shadows. We are getting increasingly good at preventing and curing many of these problems. We know how lifestyle, particularly diet and exercise, can prevent heart disease, stroke, and diabetes, as well as obesity and the problems it causes. We've talked about much of this research earlier in this book, but the punch line is straightforward and familiar: get physical activity throughout the day, every day, and find a diet you enjoy that keeps your weight in a healthy range. If you have a hard time with that (as many of us do), there's a range of effective medicines for high blood pressure, obesity, and other risk factors of heart and metabolic disease that you might want to check out with your doctor. For the first time in decades, there's reason to be optimistic about cardiovascular disease and obesity.

We've gotten much better at preventing and treating infections as well. Clean water, effective sewage treatment, and waste disposal are the cornerstones of public health. They've saved more lives than any other intervention in history. When infections do arise, we have an arsenal of effective vaccines, antibiotics, and antivirals. Their impact has been transformative. In 1900, infectious diseases were the leading killers in the U.S. Today, they don't even make the top ten. We don't do enough to tackle infections affecting poor or marginalized people—malaria remains a massive killer globally, and it took far too long to focus adequate attention to HIV, for example. But when we put our minds to it, we can develop effective treatments quickly. Vaccines and antivirals for COVID were available within months, and the mRNA technology developed for COVID vaccines has ushered in a new era in the battle against infectious disease.

Even cancer, the quintessential silent killer, has begun to show its weaknesses. We've learned that you can reduce your risk for some cancers with simple lifestyle changes. Wear sunscreen, don't smoke, take it easy with alcohol and cured meats, and you'll reduce your chances of skin, lung, liver, stomach, and colon cancer. Exercise and a healthy body weight reduce the risk of cancer in your reproductive organs such as breast, uterus, and prostate. You can't eliminate your risk of cancer completely, but you can reduce it, and recent advances are making the prospect of a positive diagnosis less terrifying.

Cancer is the price that life was willing to pay for the advantages of growth. Every time a cell divides it must duplicate its

DNA. Evolution has given us lots of mechanisms to weed out errors in the copying process, but there's a lot of DNA to copy. Austad, the longevity researcher, estimates that your cells produce *two miles* of DNA every second. There's always some chance that a DNA mutation will occur in the process, one that will cause the resulting cells to keep dividing, out of control. That nonstop cell division is what we call cancer. Cells are constantly dividing to repair and replace old tissues, and some organs are on a faster schedule than others. That's the main reason that cancer is more likely to occur in your colon, which is constantly replacing its lining, than in your bones, which are less active. The alleles you carry can also increase your risk of a bad mutation, which is why colon cancer, breast cancer, and a couple of others tend to run in families.

Regardless of where these bad divisions start, natural killer cells and cytotoxic T cells do a great job stopping them before they cause trouble. But the system isn't perfect. The good news is that we can treat many kinds of cancer that we couldn't just a few years ago. In addition to traditional approaches like surgery, chemotherapy, and radiation, researchers have figured out how to train the immune system to recognize and kill cancer cells. Two common approaches, checkpoint inhibitors and CAR T-cell therapies, modify the tools that cytotoxic T cells use to check the MHC proteins on the surfaces of other cells. Cancer cells are often sneaky, presenting MHC proteins that fool natural killer cells and cytotoxic T cells into letting them live. Checkpoint inhibitors and CAR T-cell therapies modify T cells so they aren't fooled. In CAR T-cell therapy, doctors modify your own cytotoxic T cells to recognize the cells in your specific cancer. These immunotherapies, along with other new medicines that can precisely target cancer cells, hold the promise of finally taming cancer.

Dementia, the leading cause of mortality after cardiovascular disease and cancer in the U.S. and other industrialized countries, is proving to be a more formidable opponent. We currently have no effective therapies to reverse the deterioration of brain function, including memory loss and behavioral changes, that accompany dementia. There have been some promising developments, but the current medications for Alzheimer's disease, which accounts for half of dementia deaths in the U.S., only ease the symptoms and slow progression. They don't reverse or cure the disease, and they don't work for everyone.

Your best bet is prevention, and here again a healthy lifestyle is known to help. Regular exercise is associated with clearer cognition and better brain function as people age, while hours spent sedentary on the couch or at a desk increase dementia risk. Diet studies have been less compelling, and there's no evidence that the superfoods often touted for brain function have any measurable effects. But obesity *is* a risk factor for dementia, and, as we saw in chapter 5, diet is the single biggest factor for body weight. That may be why some studies show a protective effect of Mediterranean and other whole-food-focused diets, compared to standard Western diets heavy in processed foods and added sugars and oils that push us to overeat.

Sleep is also emerging as an important lifestyle factor for dementia risk. We haven't discussed sleep very much, but it's integral to the healthy function of every system in your body. For as long as life has been evolving on our planet, the cycle of day and night has been a constant rhythm. Organisms across the tree of life, from plants to animals to single-celled creatures, have incorporated that rhythm into their biology. We see this circadian rhythm in our own bodies in the daily rise and fall of hormone levels, body temperature, mental acuity, blood pressure, and other measures. Sleep is an essential part of our daily cycle, and we tend to fall apart when we don't get enough or don't keep a regular schedule. Poor, irregular sleep puts you at risk for a long list of health problems, including Alzheimer's and other forms of dementia.

Genetics also influence your dementia risk. For Alzheimer's disease, the alleles you carry at more than eighty genes affect your chances, and one gene, *APOE*, is particularly important. There are three common alleles for *APOE* in our species, ε2, ε3, and ε4. Like most genes, you have two copies of *APOE*, meaning you carry two alleles of it, some combination of ε2, ε3, and ε4. The ε2 allele might reduce your risk, and ε3, the most common allele, is neutral and doesn't change your risk one way or the other. But the ε4 allele increases your risk substantially, particularly for the two to five percent of the population that carries two copies. If that describes you, it's imperative to stay active and watch your weight—and keep an eye out for new medications that might aid in prevention.

We might also want to look to our farming and foraging neighbors for insights on battling dementia. Work with the Tsimane population (a farming and foraging society in rural Bolivia whom we met in chapter 4) has shown remarkably low prevalence of Alzheimer's and other cognitive impairment, even in their seventies and beyond. It's not due to genetics: *APOEε4* is just as common in their population as it is in the U.S. We don't know how they keep dementia at bay, but it seems likely that their physically active lifestyle and diet low in fat and high in fiber help to keep their brains healthy as they age, just as they promote their extraordinary heart health (see chapter 4). It's possible that Tsimane adults also maintain a more regular sleep schedule, free from the distractions of streaming television and smartphones. Or perhaps, as some have argued, their greater exposure to pathogens keeps them sharp by better regulating immune system activity in the brain—the hygiene hypothesis for Alzheimer's disease.

THINGS FALL APART

Better strategies for preventing disease, and effective treatments when they strike, will go a long way toward extending *health* span, the years we enjoy free of disability and disease. But they won't necessarily raise the ceiling on maximum lifespan. Time catches up with us all, even if we stay healthy and do all the right things. Past age sixty, our organs begin to shrink—brains, livers, spleens, muscles all get smaller. Our cells slow down, which we can track through the decline in our metabolism and the energy we burn each day. Organ function falters. This is the second, much greater challenge in trying to live forever: we're evolved to wear out.

All animals are built to grow, reproduce, and age on a particular schedule, shaped by evolution. Species evolve a faster or slower pace of life depending on their particular ecology. Mice are evolved to grow fast and reproduce quickly, because waiting around too long to reach maturity puts them at risk of becoming a predator's dinner before they ever reproduce. Sparrows, on the other hand, can fly away from predators, and they grow and reproduce at a more leisurely pace.

As part of these "life history" strategies, evolution calibrates the amount of energy the body devotes to maintenance and repair versus growth and reproduction. Mice don't invest much in upkeep—what's the point, when death is ever present? Instead, they put their energy into growth and reproduction. As a result, even under ideal laboratory conditions, their bodies fall apart in a couple of years. Sparrows invest more in maintenance, and can live to be twenty years old.

Our bodies are built to last well into our seventies and beyond. The oldest person on record, Jeanne Calment, died in 1997 at 122 years old. But no one has ever approached the two hundred years that a bowhead whale might live, or the four hundred years of a

Greenland shark. Instead, the familiar, telltale signs of senescence accumulate. Our bodies stop investing in repair and maintenance, and it shows. Skin gets wrinkly, hair starts to turn gray, and our bodies slow down. Eventually, our organs fail us and we die of "natural" causes.

The rate of aging varies from person to person. Men tend to age faster and die younger than women, for instance, and as a result, two out of every three people who make it to one hundred are women.* Your genetics certainly play a role. Longevity is heritable, and having a parent or sibling who reaches old age increases your likelihood of doing so as well. Organs also vary in how quickly they age. Ovaries, for example, begin to senesce in a woman's late forties, long before other systems decline. As a result, humans are one of just a handful of species known to experience menopause, in which the ovaries stop cycling and females live a substantial portion of their lives with their reproductive careers behind them.

Menopause poses an evolutionary puzzle: Why would natural selection favor anyone living past their ability to reproduce? Why invest energy to maintain other systems once reproduction has ceased? The answer seems to be that our long, helpless childhoods give long-lived mothers an evolutionary advantage. They can live long enough to see their last child through to independence, and to help raise their grandchildren as well.

So evolution has given us grandmothers, but it also takes them away. In our golden years, investment in repair and maintenance declines, following our evolved schedule of planned obsolescence. Errors in our DNA accumulate, the ends of our chromosomes, called telomeres, grow shorter, and our machinery for making

---

* It's unclear why men die earlier than women, but testosterone seems a likely culprit. There's some indication that it's harmful for immune function, and ample evidence that it encourages dumb behavior. Men win nearly 90 percent of all Darwin Awards, given for poorly conceived actions that eliminate the winner from the gene pool.

proteins begins to falter. Cells' ability to sense nutrients in the bloodstream degrades, as do the mitochondria within them that produce ATP for energy. Cells lose their ability to divide.

In some cases, we know how to treat the symptoms of this decline. After menopause, for instance, women can take synthetic hormones to replace the estrogen and progesterone their ovaries produced when they were younger. Hormone replacement therapy increases the likelihood of blood clots and breast cancer, but many women find the benefits outweigh the risks. They feel better. Hot flashes are less frequent. The risk of osteoporosis is lower. Some studies suggest hormone replacement may reduce the risk of dementia and heart disease, too, but the evidence is mixed.

The current frontier in anti-aging science is to slow the rate of cellular decline. So far, the most effective treatments all involve slowing our cells down. We've known for over a century that calorie restriction is an effective way to extend lifespans in species from flies to mice to monkeys. Cutting the calories eaten each day (and cutting protein, in particular) activates pathways in our cells, led by the signaling molecule AMPK, that cause the cell to be more frugal and less active, producing fewer proteins and dividing less often. They also encourage autophagy, where old, broken bits of cellular machinery are recycled and used for energy or new parts. Less cell activity also means less oxidative stress, which is a by-product of metabolism.

Calorie restriction may well extend life in humans; studies are underway, but it will be decades before we know for sure. Or it might just make life *feel* longer. Consistently cutting your intake by 10 percent or more is hard, as anyone who has dieted can tell you. There are any number of methods out there you can try. Some people find it relatively easy to fast for certain periods of the day, say between morning and late afternoon, leaving their evenings to eat normally with their family and friends. Time-restricted eating and fasting don't seem to have any additional benefits beyond the reduced calorie intake, but they might be easier to follow. In some studies, but not all, people lose more weight when assigned to time-restricted eating versus reducing portion size during normal meals.

The holy grail, of course, would be a pill that simulated the effects of calorie restriction without all the starvation. A few promising miracle molecules are being studied, but none have proven to be a definitive cure for aging. Metformin, a common diabetes drug, and resveratrol, a molecule found in red wine and some other foods, stimulate the AMPK pathways and have gotten longevity researchers excited that they could slow the aging process. Metformin extends lifespans in mice, by around 6 percent in one study, but it's unclear if it has the same impact in us. Tens of millions of people take metformin to treat their diabetes, and it's certainly helpful in reducing mortality from the disease, but we haven't seen metformin users setting longevity records.

Resveratrol was the darling of anti-aging research in the early 2000s (you may recall the headlines extolling the health effects of red wine). Resveratrol stimulates molecules called sirtuins, which are part of the AMPK pathway, and has been clearly shown to extend lifespan—in yeast. But, as often happens in drug discovery, those early hopes haven't panned out. Resveratrol doesn't seem to have any effect on longevity in mice, and results from human studies are equally bleak. That hasn't stopped prominent researchers, some with commercial as well as academic interests in the drug, from pushing resveratrol as a tool to stop aging.

Rapamycin, a molecule extracted from bacteria in the soil of Easter Island, inhibits the mTOR pathway, which is the yin to the AMPK pathway's yang. If AMPK is the brake pedal, mTOR is the accelerator, ramping up protein production, growth, and cell division. It's usually hijacked in cancer, causing runaway cell division.

Rapamycin stops the mTOR pathway, with many of the same downstream effects as stimulating AMPK. Its suppressive effects on the mTOR pathway have led to the widespread use of rapamycin to aid cancer treatments and suppress immune response and tissue rejection in transplant recipients. In 2014 rapamycin was tested as a longevity drug in lab mice, and the effects were impressive, extending lifespan in some lines by around 25 percent. Similar effects in humans would raise life expectancy in the U.S. to around ninety-five years! But we don't know if it works in us. People receiving rapamycin as part of their treatment for cancer or organ transplant aren't living that long, but, of course, they might not provide a fair test of longevity effects. Trials in healthy adults are ongoing, with no clear pattern emerging yet.

Other drugs, supplements, and treatments are being explored. There's massive interest in longevity, and guaranteed fame and fortune to the lucky researchers who discover a way to fend off Father Time. As I write this, researchers at Columbia University have just published a study showing that taurine, an amino acid found in many foods, extends lifespan in mice and shows promising effects in people. There are drugs to add length back on to your telomeres, and if you're a billionaire you can even arrange to be transfused with a younger person's blood.

It's anyone's guess which, if any, of these treatments are likely to make it into your doctor's standard tool kit. But even if they do, none of them are likely to make you younger. These treatments only slow the aging process. What none of them do, and what we have yet to figure out *how* to do, is to make our cells behave like they're younger. In your twenties, your body takes care of itself, investing in repair and maintenance. As you age, those processes slow down. The best treatments we have at the moment *might* slow that decline, but none of them have shown the ability to reverse it, to point the arrow of time back toward our youth.

There's a strong argument to be made that we *never* will figure out how to stop aging, much less to age in reverse. Aging, like height, metabolism, and most other traits, is complex. We haven't identified most of the genes involved yet, but it is virtually guaranteed that thousands contribute. Changing the way our bodies age could very well require changing the ways that hundreds or thousands of those genes behave.

Height provides a useful point of reference. We can (and should) try to optimize children's nutrition and health so that they can achieve the full stature that their genes will allow. But they'll still be human-sized, not giraffe-sized. Longevity is similar. Even if we optimize our lifestyles and solve challenges like cancer and heart disease, we'll still end up with human-sized lifespans, not those of Greenland sharks. Doubling lifespan could be the equivalent of engineering people to grow twelve feet tall, requiring massive manipulations that would undoubtedly have unexpected effects. It will certainly require a major advance in our understanding of the body, a quantum leap in the science of human physiology. I suppose such an advance is possible, if we don't destroy ourselves and the planet first, but I doubt any of us will be alive to see it.

Steven Austad is more optimistic. He believes the first person to reach 150 years old is already alive today. Austad has publicly bet $1 billion on it, with Jay Olshansky, another, more skeptical longevity researcher, taking the other side of the proposition. The catch is that no one can collect until the year 2150, when the modest amount of money they put into escrow in 2000 is expected to mature to the full value of the wager. That's when Austad and Olshansky, or more likely their descendants, will know for certain whether anyone walking around today is destined to blow out the candles on their one hundred and fiftieth birthday cake.

What Olshansky and Austad agree on is the importance of

maximizing health span, not just lifespan. Living to 150 doesn't count, in real life or for the terms of their bet, if the person is too frail to hold a conversation and take care of themselves. We want our last days to be fun and full of life. And while modern medicine might extend our horizons, getting the most out of our lives will require more than a healthy diet and exercise or a miracle pill. If we want to maximize our health span, we need each other.

LIVING TOGETHER

We are a social species. From the first population of Paleolithic *Homo* that started sharing the foods they hunted and gathered, to early farming communities that worked together to plant and harvest, to our modern lives connected globally today, we have always depended on one another to survive. Isolation is so unnerving and painful that, historically, we have reserved it as a punishment for the worst offenders of societal norms and communal trust.

Our Stone Age ancestors were so tightly knit and interdependent that our bodies evolved to require social connection. Feeling lonely activates your body's fight-or-flight stress response, causing the adrenal glands that sit atop your kidneys to produce epinephrine and cortisol. Cortisol is a hormone that channels energy away from maintenance tasks like immune function and toward immediate readiness. It elevates blood glucose levels, for example, so muscles have fuel to escape a threat. Short bursts of cortisol in response to stress are a healthy part of our normal physiology. Chronic elevation is not, and it leads to serious problems. Long-term loneliness is associated with elevated blood pressure, reduced immune function (but, paradoxically, increased inflammation), and a range of mental and physical health problems. Social connectedness adds years to life expectancy and healthy years lived, particularly for older adults. Loneliness can kill you.

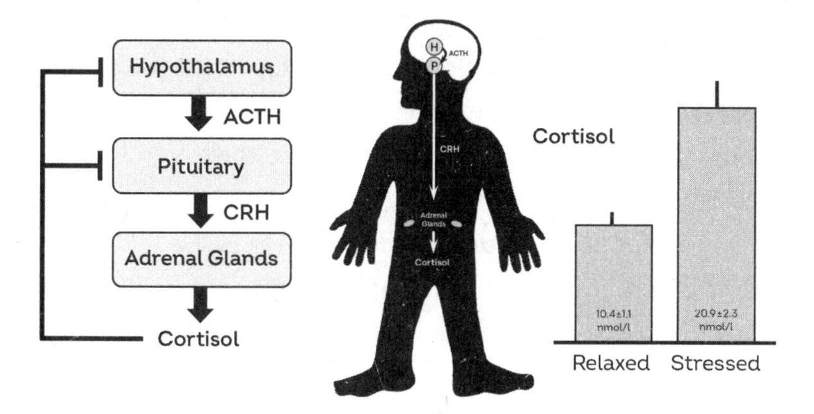

Figure 9.1 Cortisol production. The hypothalamus (H) produces ACTH, which stimulates the pituitary gland (P) to make CRH, which causes the adrenal glands to produce cortisol. Like many other hormones, cortisol is regulated through negative feedback; it inhibits the hypothalamus and pituitary, keeping its levels in check. In response to psychological stress, like the public speaking task performed in the study shown here, cortisol levels can increase by 100 percent or more.

Other social stressors can be just as potent. We are built to be part of a group, embedded within our community and surrounded by others who are on our side. When we live our lives as outsiders, bombarded every day by the message that we don't belong, that we aren't full members of the community, or that we aren't liked, our bodies suffer. Racism activates our cortisol response, with the same negative long-term consequences we see with loneliness. Immigrants can suffer the same effects. A lifetime spent swimming in cortisol is physically harmful, contributing to shorter life expectancies for Blacks, Latinos, Native Americans, and other marginalized groups in the U.S.

Social dislocation and marginalization seem to be important factors underlying "deaths of despair" from suicide, alcoholism, and overdose. Deaths of despair have risen alarmingly in the past two decades in the U.S., particularly among blue-collar and rural White communities. Easy access to prescription opioids and guns, along with potent street drugs laced with fentanyl and other deadly substances, have no doubt contributed to the problem, but the epidemic is grounded in larger societal issues that push people to abuse drugs and alcohol, or even to take their own lives. The widespread loss of manufacturing jobs and stagnation in wages that has gutted the American middle class seems a likely factor. Eastern Europe witnessed a similar epidemic in the 1990s, with the collapse of the Soviet Union and the massive economic disruption that followed. In both cases, people came to feel unvalued and aimless. The social framework they depended on dissolved.

Deaths of despair led to three consecutive years of decreasing in life expectancy in the U.S. after 2014, the first time we saw that number go the wrong way in half a century. The COVID pandemic pushed life expectancy even lower in 2020 and 2021. Forget living for 150 years. We're not even living as long as we used to.

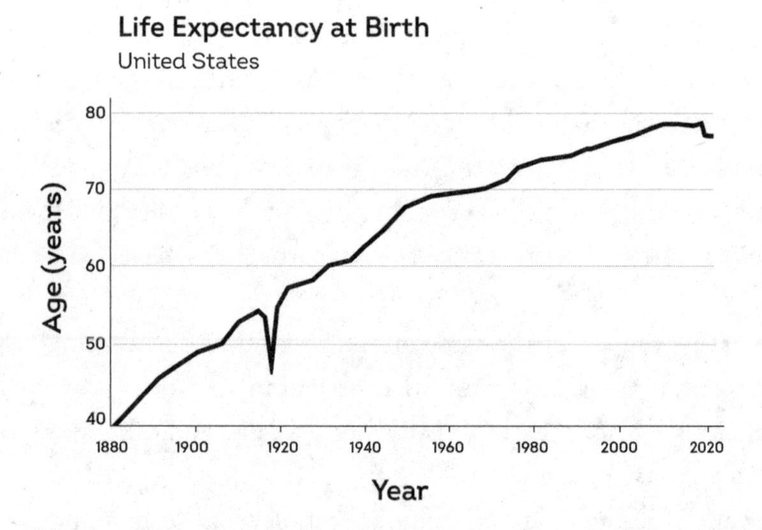

Figure 9.2 Life expectancy in the United States has nearly doubled since the 1880s. The dip in 1918 was due to the Spanish flu pandemic. The recent decline over the past few years is due to the rise in "deaths of despair" and the COVID pandemic. Life expectancy at birth for hunter-gatherer and farming communities today is around forty years, due mostly to childhood mortality from infectious disease, just as it was in the U.S. in the 1800s.

IF A PERSON WERE 150 YEARS OLD TODAY, SHE'D BE A LIVING WITNESS to remarkable advances in our understanding of the human body. Born in the aftermath of the U.S. Civil War, she would have learned as a child that Black and White people were fundamentally, biologically different. The eugenics movement would have been in full voice as she started her family in the 1890s, with experts arguing that differences in health, behavior, and intelligence among children were due to genetics.

Germ theory was still in its infancy. Pasteur and others were pushing the boundaries of microbiology and disease, but the arsenal of vaccines was tiny and effective antibiotics were decades away. Scarlet fever, easily curable today with antibiotics, was a leading cause of death among children and would have loomed large in her thoughts. In 1918, as her son and his friends came home from the ravages of the Great War in Europe, the Spanish flu pandemic swept across the globe, killing tens of millions more.

In her seventies, as she watched her relatives return home from a second world war, ideas about race and genetic determinism began to shift. The atrocities uncovered at Nazi concentration camps made the horrors of eugenics all too clear. And work at places like the Davenport orphanage made it plain just how strongly our environments shape our minds and bodies. Medicine was changing as well. Antibiotics and vaccines were saving lives. New killers, heart disease, cancers, and diabetes, were on the rise. People were beginning to wonder whether the technological miracles that brought them microwave dinners and hours in front of the television had an unexpected dark side.

Having gotten the right to vote in 1920, when she was forty-seven, she would have championed the cause of second-wave feminism in the 1960s and '70s, forcing society to rethink the relationship between sex and gender and insisting on equal rights and opportunities for women. Crucially, those changes were centered on a fuller understanding of women's bodies and abilities. Scientific breakthroughs in reproductive endocrinology gave women access to effective contraception. Soon, these would lead to modern fertility treatments like IVF. Along with abortion rights, the Pill and IVF gave women control over their bodies and their reproductive choices.

Running parallel to these social advances was a growing consensus that race is a cultural invention, not some deep, biological reality. The discovery of DNA in 1953 ushered in the era of modern genetics. For the first time, we could survey patterns of global genetic diversity. Old ideas about racial differences began to fall away. By the 1980s, it was clear that we are much more alike than we are different, and that genetic differences between populations are small and superficial. We came to understand ourselves as an African species, thoroughly mixed through global migration and intermarriage. We began to see disparities in health and educational attainment as products of racism in our society rather than racial differences in our biology.

This burgeoning understanding of our diversity and the interplay of genes and environment held broader, moral dimensions as well. Being different didn't mean you were bad or had something wrong with you. The stigma around sexual orientation and other modes of diversity began to lift. Perhaps some of her lifelong friends, ashamed to live openly in the past, came out as gay (or even left-handed).

She would have rung in the new millennium at age 127, no doubt thankful for medical advances to keep heart disease and other age-related problems in check. Hope was in the air. The Cold War was over, we had effective treatments for HIV, the internet seemed like a fun idea, and the full 3 billion letter sequence of the human genome was soon to be published. But there were clouds on the horizon. Climate change was coming into focus. Obesity and related diseases were on the rise. A new generation of opioid painkillers was being recklessly prescribed.

Alarmingly, she would have seen old, dead, and dangerous ideas coming back to life. Anti-vax conspiracy theories. Eugenics and so-called race science, reinvigorated by a misguided reading of modern genetics. Popular politicians claiming that some populations don't have "good genes," or that immigrants were "poisoning the blood of our country." People growing distrustful of scientists, and of anyone who doesn't share their skin color or worldview. Today, at 150 years old, she'd wonder if we were still moving in the right direction.

I'M OPTIMISTIC FOR THE FUTURE, DESPITE THE SIZE OF THE CHALLENGES ahead. The key lies in understanding ourselves. Knowing how our bodies work feels like turning on the lights in the middle of a scary movie. The monsters lose their heft. Yes, today's societal divisions are stark, but we now know they're born from cultural inventions that can change, not from unbridgeable biological differences. The world we've engineered can make us sick, but we know how to make it healthier. The scale of the problems ahead is enormous, but we know they're solvable.

I don't expect to make it to 150, but I suppose, if Steven Austad's correct, that my children might. Their unique mix of alleles and life experience, the opportunities that luck places in their paths and the choices they make, will determine how their lives unfold. None of us knows what scientific discovery might allow them to cheat death and keep their cells humming like they do today, full of vigor. But you can be certain that it will need to be a group effort. If we want our children to live longer and healthier lives we will need to come together, to understand ourselves and our diversity, and finally come to grips with the reality that we're all one species, our fates tied inexorably to one another. We'll need to see that our differences unite us. Only then can we cooperate at the scale needed to overcome the societal and environmental challenges that dampen our days and shorten our lives.

Life isn't a solo endeavor. It's a team project, a test of our humanity. Longevity is a measure of how well we're doing. If my kids do live to 150, it will mean we've figured out how to take care of ourselves and to look out for one another, to build a world that's less divisive and more sustainable. I hope, for the sake of our species, that they get there. And I hope they get to know Steven Austad's family. I don't know what kind of fun 150-year-olds get up to, but I imagine $1 billion will go a long way.

ACKNOWLEDGMENTS

Much of the material in this book was shaped in the classroom, teaching human physiology from an evolutionary perspective for nearly two decades. I thank my many students over the years for pushing me to see different perspectives and dig deeper into the material. People with the curiosity and courage to learn new things inevitably ask the best questions.

I'm similarly indebted to my teachers through the years, including my family—Mom, Dad, George, Heide, Holly, and Emily—and the educators I've been so incredibly lucky to learn from, from Fox Township Elementary through college and graduate school. Of particular relevance to the material here, I thank Alan Walker and Jeff Kurland for igniting my interest in human evolution, and Andrew Biewener, Peter Ellison, and Farish Jenkins for foundational lessons in the physiology and anatomy of humans and other animals.

My wife, Janice, and kids, Alex and Clara, were incredibly patient with me as I stole hours away to work on this book, and I thank them for all these years of love and support. Love you guys. Thank you.

It has been my privilege to live and work with some of the farming and foraging communities we meet in this book, and I thank them for their generosity and hospitality. My personal stories from fieldwork in those communities (and elsewhere) are true and rendered as accurately here as my memory allows. Most of these communities are under pressure as the forces of economic development threaten their land and way of life. If you're inspired to visit these groups or others, please do your homework and travel responsibly. Hire guides with a track record of working with the communities and looking out for their interests. And try to give back, if you can. There are a number of charitable organizations working with subsistence populations around the world to protect their way of life. My colleagues and I have founded one such organization to help the Hadza community, the Hadza Fund (hadzafund.org), and I invite you to check it out.

My forays into the field have all been team efforts, and I'm indebted to friends and colleagues who made all the hard work not just possible but fun, including Mariamu Anyawire, David Braun, Matt Douglas, Beatrice Eyomo, Koriye Koriye, Luke Lomeiku, Audax Mabulla, Carla Mallol, Emmanuel Ndiema, Rosemary Nzunza, Bunga Paolo, Daudi Peterson, David Raichlen, Asher Rosinger, Chris and Nani Schmeling, and Brian Wood. I'm similarly grateful to collaborative research opportunities with other communities that relied on travel and fieldwork by Samuel Urlacher, Michael Gurven, Patricia Ukegbu, and others.

Back at Duke University, I've had the great fortune of working with an incredible group of graduate students and postdocs. The current team includes Mayowa Adegboyega, Jake Diana, Elena Hinz, Lev Kolinski, Emily Hyatt, Mary Joy, Delaney Knorr, Amanda McGrosky, Srishti Sadhir, Allie Schrock, Caroline Shearer, Alma Solis, Anna Tavormina, Eric Trexler, Daniela

Trujillo, and Melody Young, as well as a great group of amazing Duke undergraduates.

My field and lab colleagues have all helped to shape my thinking on the material in this book, and many provided comments on drafts. I'm grateful for discussions over the years and comments on drafts from a long list of others as well, including Steven Austad, Danielle Belardo, Charles Brenner, Doriane Coleman, Kevin Davy, Christine Drea, David Eagle, Victoria Ehrhardt, Michael Emmanuel, Agustín Fuentes, Amy Goldberg, Elaine Guevara, Kevin Hall, Brian Hare, Nina Jablonski, Amanda Lea, Kristi Lewton, Daniel Lieberman, Jan Losos, Amy Luke, Riley MacLean, Edward Melanson, Stephanie Meredith, Hari Mix, Charlie Nunn, David Puts, Olalla Prado Nóvoa, Eric Ravussin, Leanne Redman, Charles Roseman, Dale Schoeller, Jay Schoenherr, Pat Shipman, John Speakman, Anne Stone, Deirdre Thomas, Ben Trumble, Ethan Weiss, Klaas Westerterp, Sarahn Wheeler, Chris Wildeman, Vanessa Woods, Yosuke Yamada, and Guillermo Zorrilla. Thank you all for your time and expertise.

I thank Duke University, the Department of Evolutionary Anthropology, the Duke Global Health Institute, and the Duke University Population Research Institute for their support. In particular, Lisa Jones, Denise Hayes, and Kinley Johnson keep the ship afloat—thank you! I'm also grateful to the organizations that have funded my research over the years, including studies that are highlighted in this book: the National Science Foundation, National Institutes of Health, Bill and Melinda Gates Foundation, Leakey Foundation, and Wenner-Gren Foundation.

I thank my agent, Max Brockman, for his advice and guidance, and editors Hannah Steigmeyer and Caroline Sutton and the production team at Avery and Penguin Random House for shepherding this project from concept to publication.

Finally, a note of thanks and love to my community of friends here in North Carolina, my adopted forever home. From Pod Dinners to Climbing Wednesdays, beers and darts to mountain trips and beach weekends, y'all keep me sane and motivated and grateful for this one incredible, crazy trip we get to take through life. May your respective protein robots serve you well.

NOTES

Foundational details about the anatomy and physiology of various systems (e.g., the structure of the heart, or the regulation of reproductive hormones) can be found in a college-level human physiology textbook. Sources for other research covered in this book are given below.

INTRODUCTION: DISASSEMBLY REQUIRED

xiii **MEDICAL PROFESSIONALS . . . SHARE MANY OF THE WIDESPREAD AND UNFOUNDED BIASES:** Kelly M. Hoffman et al., "Racial Bias in Pain Assessment and Treatment Recommendations, and False Beliefs about Biological Differences between Blacks and Whites," *Proceedings of the National Academy of Sciences* 113, no. 16 (April 19, 2016): 4296–301; Chuck Galli and Tiffany Li, "Racial Differences in Diagnosis of Overweight and Obesity: Results from the National Health and Nutrition Examination Survey (NHANES), 2009–2016," *Journal of Racial and Ethnic Health Disparities* 10, no. 3 (April 8, 2022): 1096–107.

xiv **ALGORITHMS THAT DOCTORS RELY ON EVERY DAY:** Darshali A. Vyas, Leo G. Eisenstein, and David S. Jones, "Hidden in Plain Sight—Reconsidering the Use of Race Correction in Clinical Algorithms," *New England Journal of Medicine* 383, no. 9 (August 27, 2020): 874–82.

xvi **EVERY ONE OF US ALL OVER THE WORLD IS MORE THAN 99.9 PERCENT SIMILAR IN OUR DNA:** "Genetics versus Genomics Factsheet," National Human

Genome Research Institute, accessed February 1, 2024, https://www.genome.gov/about-genomics/fact-sheets/Genetics-vs-Genomics.

CHAPTER 1: SO SIMPLE A BEGINNING

5 **PRODUCTION OF NEURONS . . . BEGINS IN THE SIXTH WEEK AFTER FERTILIZATION AND CONTINUES THROUGHOUT GESTATION:** Joan Stiles and Terry L. Jernigan, "The Basics of Brain Development," *Neuropsychology Revue* 20, no. 4 (December 2010): 327–48.

6 **ROUGHLY 30 PERCENT OF EMBRYOS ARE SPONTANEOUSLY ABORTED:** M. J. Zinaman et al., "Estimates of Human Fertility and Pregnancy Loss," *Fertility and Sterility* 65, no. 3 (March 1996): 503–9.

9 **THOSE WHO BELIEVE THAT LIFE BEGINS AT CONCEPTION:** Patrick Lee and Robert P. George, "Embryology, Philosophy & Human Dignity," *National Review*, August 9, 2001.

9 **PETER SINGER HAS FAMOUSLY ARGUED:** From his website and writings: https://petersinger.info/faq.

10 **CRANIOPAGUS PARASITICUS:** "Craniopagus Parasiticus," Wikipedia, accessed February 1, 2024, https://en.wikipedia.org/wiki/Craniopagus_parasiticus.

11 **CONNECTIONS BETWEEN NEURONS NEEDED FOR NORMAL BRAIN FUNCTION DON'T DEVELOP MUCH BEFORE TWENTY-TWO WEEKS:** András Jakab et al., "Fetal Functional Imaging Portrays Heterogeneous Development of Emerging Human Brain Networks," *Frontiers in Human Neuroscience* 8 (October 22, 2014): 852.

11 **SOME ARGUE IT'S NOT EVEN THE CENTRAL QUESTION:** Liberal in a Red State, "It Doesn't Matter When 'Life' Begins, No Person Can Force You to Use Your Body to Save Their Life," *Daily Kos*, May 22, 2022.

12 **YOU'RE ALLOWED TO CHANGE YOUR MIND:** "Making the Decision to Donate," National Kidney Foundation, accessed February 13, 2024, https://www.kidney.org/transplantation/livingdonors/making-decision-to-donate.

13 **EVERY CULTURE HAS ITS OWN LIST OF DOS AND DON'TS DURING PREGNANCY:** China: Ying Lau, "Traditional Chinese Pregnancy Restrictions, Health-Related Quality of Life and Perceived Stress among Pregnant Women in Macao, China," *Asian Nursing Research* 6, no. 1 (March 2012): 27–34. Zulu: Mmbulaheni Ramulondi, Helene de Wet, and Nuntuthuko Rosemary Ntuli, "Traditional Food Taboos and Practices during Pregnancy, Postpartum Recovery, and Infant Care of Zulu Women in Northern KwaZulu-Natal," *Journal of Ethnobiology and Ethnomedicine* 17, no. 1 (March 20, 2021): 15. Ghana: Joana Ansong, Emmanuel Asampong, and Philip Baba Adongo, "Sociocultural Beliefs and Practices during Pregnancy, Child Birth, and Postnatal Period: A Qualitative Study in Southern

Ghana," *Cogent Public Health* 9, no. 1 (2020): 2046908. U.S.: Kerry Dougherty, "Q&A: Is It Safe to Raise My Arms above My Head While Pregnant?," Bump, March 2, 2017, https://www.thebump.com/a/is-it-safe-to-raise-arms-above-head.

13 **YOU CAN STILL DRINK COFFEE AND TEA:** Stefanie N. Hinkle et al., "Assessment of Caffeine Consumption and Maternal Cardiometabolic Pregnancy Complications," *JAMA Network Open* 4, no. 11 (November 1, 2021): e2133401.

14 **DUTCH FAMINE OF 1944-45:** "Dutch Famine of 1944–1945," Wikipedia, accessed February 1, 2024, https://en.wikipedia.org/wiki/Dutch_famine_of_1944-1945.

14 **A TYPICAL PREGNANCY FOR A HEALTHY WOMAN IN A WELL-NOURISHED POPULATION REQUIRES OVER SEVENTY-FIVE THOUSAND KILOCALORIES:** Nancy F. Butte and Janice C. King, "Energy Requirements during Pregnancy and Lactation," *Public Health Nutrition* 8, no. 7A (October 2005): 1010–27.

14 **MEDICAL RECORDS OF MOTHERS WHO WERE PREGNANT DURING THE FAMINE:** Laura S. Bleker et al., "Cohort Profile: The Dutch Famine Birth Cohort (DFBC)—A Prospective Birth Cohort Study in the Netherlands," *BMJ Open* 11, no. 3 (March 4, 2021): e042078.

15 **DAVID BARKER PUBLISHED LANDMARK STUDIES:** D. J. Barker and C. Osmond, "Infant Mortality, Childhood Nutrition, and Ischaemic Heart Disease in England and Wales," *Lancet* 1, no. 8489 (May 10, 1986): 1077–81; D. J. Barker et al., "Weight in Infancy and Death from Ischaemic Heart Disease," *Lancet* 2, no. 8663 (September 9, 1989): 577–80.

15 **MORE LIKELY TO DEVELOP HEART DISEASE, DIABETES, AND OTHER PROBLEMS:** D. J. P. Barker, "The Developmental Origins of Adult Disease," *Journal of the American College of Nutrition* 23, suppl. 6 (2004): 588S–595S.

15 **SEE THIS EFFECT WITH DUTCH FAMINE BABIES AS WELL:** Bleker et al., "Cohort Profile: The Dutch Famine Birth Cohort (DFBC)."

16 **CURRENT THINKING FOCUSES ON TWO EFFECTS:** M. Desai, J. K. Jellyman, and M. G. Ross, "Epigenomics, Gestational Programming and Risk of Metabolic Syndrome," *International Journal of Obesity* 39, no. 4 (April 2015): 633–41.

16 **DUTCH FAMINE BABIES, EPIGENETIC CHANGES:** Elmar W. Tobi et al., "DNA Methylation Differences after Exposure to Prenatal Famine Are Common and Timing- and Sex-Specific," *Human Molecular Genetics* 18, no. 21 (November 1, 2009): 4046–53; Bastiaan T. Heijmans et al., "Persistent Epigenetic Differences Associated with Prenatal Exposure to Famine in Humans," *Proceedings of the National Academy of Sciences* 105, no. 44 (November 4, 2008): 17046–49.

17 **PASS THOSE EFFECTS TO THEIR CHILDREN AND *GRANDCHILDREN*:** G. Kaati, L. O. Bygren, and S. Edvinsson, "Cardiovascular and Diabetes Mortality Determined by Nutrition during Parents' and Grandparents' Slow Growth

Period," *European Journal of Human Genetics* 10, no. 11 (November 2002): 682–88; Stephanie E. King and Michael K. Skinner, "Epigenetic Transgenerational Inheritance of Obesity Susceptibility," *Trends in Endocrinology and Metabolism* 31, no. 7 (July 2020): 478–94.

CHAPTER 2: GROWING UP

22 **YOUR BONES BEGIN TO FORM IN THE SIXTH AND SEVENTH WEEKS AFTER CONCEPTION:** Grant Breeland, Margaret A. Sinkler, and Ritesh G. Menezes, *Embryology, Bone Ossification* (Treasure Island, FL: StatPearls, 2020).

24 **SMALLER JAWS CHANGED THE WAY WE TALK:** D. E. Blasi et al., "Human Sound Systems Are Shaped by Post-Neolithic Changes in Bite Configuration," *Science* 363, no. 6432 (March 15, 2019): eaav3218.

27 **U.S. PRESIDENTS HAVE ALL BEEN TALLER THAN AVERAGE:** "Heights of Presidents and Presidential Candidates of the United States," Wikipedia, accessed February 1, 2024, https://en.wikipedia.org/wiki/Heights_of_presidents_and_presidential_candidates_of_the_United_States.

28 **THE BENEFITS OF BEING TALL:** Angus Deaton and Raksha Arora, "Life at the Top: The Benefits of Height," *Economics and Human Biology* 7, no. 2 (July 2009): 133–36; Anne Case and Christina Paxson, "Stature and Status: Height, Ability, and Labor Market Outcomes," *Journal of Political Economy* 116, no. 3 (June 2008): 499–532; Daniel LaFave and Duncan Thomas, "Height and Cognition at Work: Labor Market Productivity in a Low Income Setting," *Economics and Human Biology* 25 (May 2017): 52–64.

28 **THE DATING SITE OKCUPID:** "The Big Lies People Tell in Online Dating," *Medium,* July 7, 2010, https://theblog.okcupid.com/the-big-lies-people-tell-in-online-dating-a9e3990d6ae2.

29 **DNA IS A THREADLIKE MOLECULE:** "Essentials of Genetics: Unit 1: What Is DNA? What Does DNA Do?," Nature Education, accessed February 1, 2024, https://www.nature.com/scitable/ebooks/essentials-of-genetics-8 /126042179/.

31 **HERITABILITY VALUES FOR HEIGHT ARE TYPICALLY AROUND 0.50 TO 0.70:** Loïc Yengo et al., "A Saturated Map of Common Genetic Variants Associated with Human Height," *Nature* 610, no. 7933 (October 2022): 704–12.

32 **OVER TWELVE THOUSAND SNPS ASSOCIATED WITH HEIGHT:** Yengo et al., "A Saturated Map of Common Genetic Variants Associated with Human Height."

33 **MORE THAN TWO HUNDRED GENES HAVE BEEN IDENTIFIED THAT CONTRIBUTE TO FACIAL SHAPE:** Julie D. White et al., "Insights into the Genetic Architecture of the Human Face," *Nature Genetics* 53, no. 1 (January 2021): 45–53.

33 **THESE GENES ARE WHY PEOPLE FROM DIFFERENT PARTS OF THE WORLD HAVE DISTINCTIVE FACIAL FEATURES:** M. Zhang et al., "Genetic Variants

Underlying Differences in Facial Morphology in East Asian and European Populations," *Nature Genetics* 54, no. 4 (April 2022): 403–11.

37 **GROW TALLER THAN THOSE WHO REMAIN IN THEIR HOMELAND:** Barry Bogin, Michael Hermanussen, and Christiane Scheffler, "As Tall as My Peers—Similarity in Body Height between Migrants and Hosts," *Anthropologischer Anzeiger* 74, no. 5 (June 11, 2018): 365–76.

37 **KIDS SENT TO WORK IN FACTORIES WERE ABOUT FOUR INCHES (TEN CENTIMETERS) SHORTER:** J. M. Tanner, "Growth as a Mirror of the Condition of Society: Secular Trends and Class Distinctions," *Acta Paediatrica Japonica* 29, no. 1 (February 1987): 96–103.

38 **FARMING GENERALLY PRODUCES MORE FOOD FOR LESS EFFORT THAN HUNTING AND GATHERING:** Thomas S. Kraft et al., "The Energetics of Uniquely Human Subsistence Strategies," *Science* 374, no. 6575 (December 24, 2021): eabf0130.

38 **FERTILITY INCREASED, BUT SO DID CHILDHOOD MORTALITY AND DISEASE:** Jean-Pierre Bocquet-Appel, "When the World's Population Took Off: The Springboard of the Neolithic Demographic Transition," *Science* 333, no. 6042 (July 29, 2011): 560–61.

38 **THE SHUAR POPULATION:** "Shuar," Wikipedia, accessed February 1, 2024, https://en.wikipedia.org/wiki/Shuar.

38 **WORK LED BY ANTHROPOLOGIST SAM URLACHER:** Samuel S. Urlacher et al., "Tradeoffs between Immune Function and Childhood Growth among Amazonian Forager-Horticulturalists," *Proceedings of the National Academy of Sciences* 115, no. 17 (April 24, 2018): E3914-E3921.

39 **KIDS ARE GROWING TALLER:** NCD Risk Factor Collaboration (NCD-RisC), "Height and Body-Mass Index Trajectories of School-Aged Children and Adolescents from 1985 to 2019 in 200 Countries and Territories: A Pooled Analysis of 2,181 Population-Based Studies with 65 Million Participants," *Lancet* 396, no. 10261 (Novemeber 7, 2020): 1511–24.

39 **THIS GROWTH HAS PLATEAUED:** Yvonne Schönbeck et al., "The World's Tallest Nation Has Stopped Growing Taller: The Height of Dutch Children from 1955 to 2009," *Pediatric Research* 73, no. 3 (March 2013): 371–77.

41 **THERE'S A "BEAUTY PREMIUM":** Daniel S. Hamermesh, *Beauty Pays: Why Attractive People Are More Successful* (Princeton, NJ: Princeton University Press, 2011); Qingguo Ma et al., "The Undermining Effect of Facial Attractiveness on Brain Responses to Fairness in the Ultimatum Game: An ERP Study," *Frontiers in Neuroscience* 9 (March 9, 2015): 77.

41 **A MORE PROVOCATIVE EXPLANATION FOR THE HEIGHT PREMIUM:** Case and Paxson, "Stature and Status."

42 **THE $530 BILLION COSMETIC INDUSTRY:** P. N. Danziger, "6 Trends Shaping the Future of the $532B Beauty Business," *Forbes*, September 1, 2019.

44 **SOLUTION TO THE PUZZLE OF DAASANACH GROWTH:** Zane S. Swanson et al.,

"Early Childhood Growth in Daasanach Pastoralists of Northern Kenya: Distinct Patterns of Faltering in Linear Growth and Weight Gain," *American Journal of Human Biology* 35, no. 4 (April 2023): e23842.

45 **SO MUCH GENETIC DIVERSITY WITHIN AFRICA:** Lynn B. Jorde and Stephen P. Wooding, "Genetic Variation, Classification and 'Race,'" *Nature Genetics* 36, suppl. 11 (November 2004): S28–S33; Pavel Duda and Jan Zrzavý, "Human Population History Revealed by a Supertree Approach," *Scientific Reports* 6, no. 2989020 (July 19, 2016): doi: 10.1038/srep29890.

CHAPTER 3: HEAD QUARTERS

52 **BRAIN HAS 86 BILLION NEURONS:** Suzana Herculano-Houzel, "The Human Brain in Numbers: A Linearly Scaled-Up Primate Brain," *Frontiers in Human Neuroscience* 3 (November 9, 2009): 31.

54 **SPEEDS IN EXCESS OF 250 MILES PER HOUR:** "Action Potential," Wikipedia, accessed February 1, 2024, https://en.wikipedia.org/wiki/Action_potential.

55 **NICOTINE FROM TOBACCO BINDS TO NEURONS AND INCREASES THE PRODUCTION OF DOPAMINE:** Neal L. Benowitz, "Pharmacology of Nicotine: Addiction, Smoking-Induced Disease, and Therapeutics," *Annual Review of Pharmacology and Toxicology* 49 (2009): 57–71.

58 **BURN MORE CALORIES EACH DAY THAN ANY OTHER ORGAN:** Christopher W. Kuzawa et al., "Metabolic Costs and Evolutionary Implications of Human Brain Development," *Proceedings of the National Academy of Sciences* 111, no. 36 (September 9, 2014): 13010–15.

58 **CATS . . . LOST THE ABILITY TO EVEN TASTE SUGAR:** Xia Li et al., "Cats Lack a Sweet Taste Receptor," *Journal of Nutrition* 136, suppl. 7 (July 2006): 1932S–1934S.

58 **MICE ARE BORN WITH A PREWIRED FEAR RESPONSE TO THE SMELL OF CAT URINE:** Ajai Vyas et al., "Behavioral Changes Induced by Toxoplasma Infection of Rodents Are Highly Specific to Aversion of Cat Odors," *Proceedings of the National Academy of Sciences* 104, no. 15 (April 10, 2007): 6442–47.

59 **CAVE-DWELLING FISH:** Damian Moran, Rowan Softley, and Eric J. Warrant, "The Energetic Cost of Vision and the Evolution of Eyeless Mexican Cavefish," *Science Advances* 1, no. 8 (September 11, 2015): e1500363.

59 **BATS EVOLVED EXPANDED BRAIN REGIONS FOR TONGUE CONTROL:** Andrew C. Halley et al., "Coevolution of Motor Cortex and Behavioral Specializations Associated with Flight and Echolocation in Bats," *Current Biology* 32, no. 13 (July 11, 2022): 2935–41.e3.

62 **LEWIS TERMAN, A EUGENICS ZEALOT:** "Lewis Terman," Wikipedia, accessed February 1, 2024, https://en.wikipedia.org/wiki/Lewis_Terman.

63 **TENS OF THOUSANDS OF PEOPLE WERE FORCIBLY STERILIZED IN THE U.S.:** "Eugenics in the United States," Wikipedia, accessed February 1, 2024, https://en.wikipedia.org/wiki/Eugenics_in_the_United_States.

64 **PROMINENT BLACK SCHOLARS, W. E. B. DU BOIS AMONG THEM:** Marilyn M. Singleton, "The 'Science' of Eugenics: America's Moral Detour," *Journal of American Physicians and Surgeons* 19, no. 4 (Winter 2014): 122–26.

65 **SKODAK AND SKEELS PLACED A GROUP OF INFANTS:** Harold M. Skeels, "Adult Status of Children with Contrasting Early Life Experiences: A Follow-Up Study," *Monographs of the Society for Research in Child Development* 31, no. 3 (1966): 1–65.

68 **HAND-WRINGING ABOUT THE FERTILITY OF BLACK, LATINO, AND OTHER MINORITY COMMUNITIES:** Edna Bonhomme, "How the Myth of Black Hyper-Fertility Harms Us," Al Jazeera, August 16, 2020; Emilio A. Parrado and Chenoa A. Flippen, "Hispanic Fertility, Immigration, and Race in the Twenty-First Century," *Race and Social Problems* 4, no. 1 (April 1, 2012): 18–30.

68 **EDUCATIONAL EXPERIENCE AND OPPORTUNITY STILL DIFFER AMONG RACIAL GROUPS:** "K–12 Disparity Facts and Statistics," United Negro College Fund, accessed February 13, 2024, https://uncf.org/pages/k-12-disparity-facts-and-stats.

68 **BLACK CHILDREN ARE TYPICALLY SEVERAL MONTHS BEHIND THEIR WHITE PEERS IN READING AND MATH:** Allison Friedman-Krauss and W. Steven Barnett, "Special Report: Access to High-Quality Early Education and Racial Equity," National Institute for Early Education Research, Rutgers Graduate School of Education, June 16, 2020, https://nieer.org/research-library/special-report-access-high-quality-early-education-racial-equity.

69 **THE FLYNN EFFECT:** "Flynn Effect," Wikipedia, accessed February 1, 2024, https://en.wikipedia.org/wiki/Flynn_effect.

70 **THE EFFECT APPEARS TO BE MODEST:** Jeanne E. Savage et al., "Genome-Wide Association Meta-Analysis in 269,867 Individuals Identifies New Genetic and Functional Links to Intelligence," *Nature Genetics* 50, no. 7 (July 2018): 912–19.

71 **BIG FIVE PERSONALITY TRAITS:** S. Sanchez-Roige et al., "The Genetics of Human Personality," *Genes, Brain and Behavior* 17, no. 3 (March 2018): e12439.

71 **PERSONALITY TRAITS AND VARIATION IN SYNAPSE AND NEUROTRANSMITTER FUNCTION:** Szymon Zmorzyński et al., "Personality Traits and Polymorphisms of Genes Coding Neurotransmitter Receptors or Transporters: Review of Single Gene and Genome-Wide Association Studies," *Annals of General Psychiatry* 20, no. 1 (January 22, 2021): 7.

71 **ALLELES ASSOCIATED WITH AGREEABLENESS AND NEUROTICISM ARE ALSO LINKED WITH GAMBLING ADDICTION:** Kellyn M. Spychala et al., "Predicting Disordered Gambling across Adolescence and Young Adulthood from

Polygenic Contributions to Big 5 Personality Traits in a UK Birth Cohort," *Addiction* 117, no. 3 (March 2022): 690–700.

71 **MUSICAL ABILITY OR ATHLETIC TALENT, ARE INFLUENCED BY THE ALLELES WE CARRY:** Fatima Al-Khelaifi et al., "Metabolic GWAS of Elite Athletes Reveals Novel Genetically-Influenced Metabolites Associated with Athletic Performance," *Scientific Reports* 9, no. 1 (December 27, 2019): 19889; Yi Ting Tan et al., "The Genetic Basis of Music Ability," *Frontiers in Psychology* 5 (June 27, 2014): 658.

74 **THE GENETIC ASSOCIATION OF IQ AND EDUCATIONAL ATTAINMENT WITH OTHER, SEEMINGLY UNRELATED TRAITS:** Abdel Abdellaoui and Karin J. H. Verweij, "Dissecting Polygenic Signals from Genome-Wide Association Studies on Human Behaviour," *Nature Human Behaviour* 5, no. 6 (June 2021): 686–94.

CHAPTER 4: HEART AND AIR SUPPLY

82 **HEART DISEASE, DIABETES, AND OTHER MISMATCH DISEASES WERE MINOR PLAYERS IN PUBLIC HEALTH:** David S. Jones, Scott H. Podolsky, Jeremy A. Greene, "The Burden of Disease and the Changing Task of Medicine," *New England Journal of Medicine* 366, no. 25 (June 21, 2012): 2333–38.

82 **THE AVERAGE AMERICAN LIVES TWO DECADES LONGER THAN THEY DID JUST A CENTURY AGO:** Saloni Dattani et al., "Life Expectancy," Our World in Data, accessed February 13, 2024, https://ourworldindata.org/life-expectancy.

82 **OVER 2 BILLION HEARTBEATS AND HALF A BILLION BREATHS BY THE TIME WE REACH OUR SEVENTIES:** Assuming, conservatively, that your heart rate averages sixty beats per minute and your respiration rate is fifteen breaths per minute.

83 **WALLS OF CAPILLARIES AND ALVEOLI ARE EACH JUST ONE CELL THICK:** S. Tsunoda et al., "Lung Volume, Thickness of Alveolar Walls, and Microscopic Anisotropy of Expansion," *Respiration Physiology* 22, no. 3 (December 1974): 285–96.

85 **GALEN OF PERGAMON:** "Galen," Wikipedia, accessed February 1, 2024, https://en.wikipedia.org/wiki/Galen.

86 **WILLIAM HARVEY:** "William Harvey," Wikipedia, accessed February 1, 2024, https://en.wikipedia.org/wiki/William_Harvey.

87 **THE SA NODE:** "Arthur Keith," Wikipedia, accessed February 1, 2024, https://en.wikipedia.org/wiki/Arthur_Keith; "Piltdown Man," Wikipedia, accessed February 1, 2024, https://en.wikipedia.org/wiki/Piltdown_Man.

92 **NATIVE POPULATIONS IN THE ANDES:** Colleen G. Julian and Lorna G. Moore, "Human Genetic Adaptation to High Altitude: Evidence from the Andes," *Genes* 10, no. 2 (February 15, 2019): 150.

94 **A PALEOLITHIC TRYST WITH A GROUP CALLED THE DENISOVANS:** Xinjun Zhang et al., "The History and Evolution of the Denisovan-*EPAS1* Haplotype in Tibetans," *Proceedings of the National Academy of Sciences* 118, no. 22 (June 1, 2021): e2020803118.

94 **A POPULATION KNOWN AS THE SAMA:** Melissa A. Ilardo et al., "Physiological and Genetic Adaptations to Diving in Sea Nomads," *Cell* 173, no. 3 (April 19, 2018): 569–80.e15.

95 **HAVE LARGER SPLEENS THAN LOWLANDERS:** Pontus Holmström et al., "Spleen Size and Function in Sherpa Living High, Sherpa Living Low and Nepalese Lowlanders," *Frontiers in Physiology* 11 (June 29, 2020): 647.

96 **SPEND MORE OF THEIR ENERGY WORKING TO GET FOOD, THAN ANY OF THE OTHER APES:** Thomas S. Kraft et al., "The Energetics of Uniquely Human Subsistence Strategies," *Science* 374, no. 6575 (December 24, 2021): eabf0130.

96 **MORE FATIGUE-RESISTANT "SLOW-TWITCH" MUSCLES IN OUR LEGS AND GREATER VO_2MAX:** Herman Pontzer, "Economy and Endurance in Human Evolution," *Current Biology* 27, no. 12 (June 19, 2017): R613–R621.

96 **OUR HEARTS ARE BUILT FOR ENDURANCE:** Robert E. Shave et al., "Selection of Endurance Capabilities and the Trade-Off between Pressure and Volume in the Evolution of the Human Heart," *Proceedings of the National Academy of Sciences* 116, no. 40 (October 1, 2019): 19905–910.

97 **MORE AMERICANS DIED FROM HEART DISEASE:** Elizabeth G. Nabel and Eugene Braunwald, "A Tale of Coronary Artery Disease and Myocardial Infarction," *New England Journal of Medicine* 366, no. 1 (January 5, 2012): 54–63.

98 **TWO RISK FACTORS: BLOOD PRESSURE AND CHOLESTEROL:** W. B. Kannel et al., "Factors of Risk in the Development of Coronary Heart Disease—Six Year Follow-Up Experience. The Framingham Study," *Annals of Internal Medicine* 55 (July 1961): 33–50.

98 **JERRY MORRIS AND HIS TEAM IN ANOTHER FIRST-OF-ITS-KIND STUDY:** J. N. Morris et al., "Coronary Heart-Disease and Physical Activity of Work," *Lancet* 262, no. 6795 (November 21, 1953): 1053–57.

99 **CHOLESTEROL IS GENERALLY BAD, TOO, PARTICULARLY THE "LDL" VARIETY:** Brian A. Ference et al., "Low-Density Lipoproteins Cause Atherosclerotic Cardiovascular Disease. 1. Evidence from Genetic, Epidemiologic, and Clinical Studies. A Consensus Statement from the European Atherosclerosis Society Consensus Panel," *European Heart Journal* 38, no. 32 (August 21, 2017): 2459–72.

99 **PHYSICAL ACTIVITY IS *REALLY* GOOD FOR YOU:** Leonardo Garcia et al., "Non-Occupational Physical Activity and Risk of Cardiovascular Disease, Cancer and Mortality Outcomes: A Dose-Response Meta-Analysis of Large Prospective Studies," *British Journal of Sports Medicine* 57, no. 15 (February 28, 2023): 979–89.

100 **MEN AND WOMEN TYPICALLY HAVE TOTAL CHOLESTEROLS AROUND 110, WITH LDL LEVELS BELOW 70:** H. Pontzer, B. M. Wood, and D. A. Raichlen, "Hunter-Gatherers as Models in Public Health," *Obesity Reviews* 19, suppl. 1 (December 2018): 24–35.

101 **CALCIFIED PLAQUES IN THE CORONARY ARTERIES OF OVER SEVEN HUNDRED TSIMANE ADULTS:** Hillard Kaplan et al., "Coronary Atherosclerosis in Indigenous South American Tsimane: A Cross-Sectional Cohort Study," *Lancet* 389, no. 10080 (April 29, 2017): 1730–39.

101 **HAVE GREATLY REDUCED HEART DISEASE'S TOLL:** Nabel and Braunwald, "A Tale of Coronary Artery Disease and Myocardial Infarction."

102 **BLACK, HISPANIC, AND NATIVE AMERICAN ADULTS ARE MUCH MORE LIKELY TO DEVELOP AND DIE FROM HEART DISEASE:** Zulqarnain Javed et al., "Race, Racism, and Cardiovascular Health: Applying a Social Determinants of Health Framework to Racial/Ethnic Disparities in Cardiovascular Disease," *Circulation: Cardiovascular Quality and Outcomes* 15, no. 1 (January 2022): e007917.

102 **BLACK AMERICANS, ARE JUST GENETICALLY SUSCEPTIBLE TO HEART DISEASE:** Richard S. Cooper et al., "An International Comparative Study of Blood Pressure in Populations of European vs. African Descent," *BMC Medicine* 3 (January 5, 2005): 2.

102 **ALLELES THAT INFLUENCE YOUR RISK OF DEVELOPING HEART DISEASE:** Zhifen Chen and Heribert Schunkert, "Genetics of Coronary Artery Disease in the Post-GWAS Era," *Journal of Internal Medicine* 290, no. 5 (November 2021): 980–92.

103 ***ABO* BLOOD TYPE ALLELES:** "ABO Blood Group System," Wikipedia, accessed February 1, 2024, https://en.wikipedia.org/wiki/ABO_blood_group_system.

104 ***A* OR *B* BLOOD TYPE ALLELES INCREASES HEART DISEASE RISK SLIGHTLY:** Zhuo Chen et al., "ABO Blood Group System and the Coronary Artery Disease: An Updated Systematic Review and Meta-Analysis," *Scientific Reports* 6 (March 18, 2016): 23250.

105 **SICKLE CELL ALLELE DOESN'T AFFECT YOUR LIKELIHOOD OF DEVELOPING HEART DISEASE OR HAVING A STROKE:** Hyacinth I. Hyacinth et al., "Association of Sickle Cell Trait with Ischemic Stroke among African Americans: A Meta-Analysis," *JAMA Neurology* 75, no. 7 (July 1, 2018): 802–7; Natalie A. Bello et al., "Sickle Cell Trait Is Not Associated with an Increased Risk of Heart Failure or Abnormalities of Cardiac Structure and Function," *Blood* 129, no. 6 (February 9, 2017): 799–801.

106 **BLACK AMERICANS WHO EXPERIENCE MORE RACISM TEND TO HAVE INCREASED LEVELS OF INFLAMMATION:** Tené T. Lewis et al., "Self-Reported Experiences of Everyday Discrimination Are Associated with Elevated C-Reactive Protein Levels in Older African-American Adults," *Brain, Behavior, and Immunity* 24, no. 3 (March 2010): 438–43.

106 **A LONG-TERM STUDY OF FORTY-EIGHT THOUSAND BLACK WOMEN:** Shanshan Sheehy et al., "Perceived Interpersonal Racism in Relation to Incident Coronary Heart Disease among Black Women," *Circulation* 149, no. 7 (February 13, 2024): 521–28.

106 **IN NIGERIA, AVERAGE BLOOD PRESSURE IS SIMILAR TO THAT OF WHITE AMERICANS:** Cooper et al., "An International Comparative Study of Blood Pressure in Populations of European vs. African Descent."

106 **YOU ARE TWENTY TIMES MORE LIKELY TO DIE FROM LUNG CANCER:** Michael J. Thun et al., "50-Year Trends in Smoking-Related Mortality in the United States," *New England Journal of Medicine* 368, no. 4 (January 24, 2013): 351–64.

107 **CHRONIC STRESS IS AN IMPORTANT RISK FACTOR FOR HEART DISEASE:** Mika Kivimäki and Andrew Steptoe, "Effects of Stress on the Development and Progression of Cardiovascular Disease," *Nature Reviews Cardiology* 15, no. 4 (April 2018): 215–29.

107 **THOSE RANDOMLY ASSIGNED TO A MEDITATION GROUP REPORTED LESS ANGER:** Robert H. Schneider et al., "Stress Reduction in the Secondary Prevention of Cardiovascular Disease: Randomized, Controlled Trial of Transcendental Meditation and Health Education in Blacks," *Circulation: Cardiovascular Quality and Outcomes* 5, no. 6 (November 2012): 750–58.

108 **AMERICAN HEART ASSOCIATION'S PREVENT RISK CALCULATOR:** PREVENT™ Online Calculator" American Heart Association. Accessed June 12, 2024. https://professional.heart.org/en/guidelines-and-statements/prevent-calculator.

109 **BASE JUMPERS HAVE AN ACCIDENT ABOUT EVERY 230 JUMPS:** Kjetil Soreide, Christian Lycke Ellingsen, and Vibeke Knutson, "How Dangerous Is BASE Jumping? An Analysis of Adverse Events in 20,850 Jumps from the Kjerag Massif, Norway," *Journal of Trauma and Acute Care Surgery* 62, no. 5 (May 2007): 1113–17.

110 **MODERN BLOOD PRESSURE MEDS:** Jingkai Wei et al., "Comparison of Cardiovascular Events among Users of Different Classes of Antihypertension Medications: A Systematic Review and Network Meta-Analysis," *JAMA Network Open* 3, no. 2 (February 5, 2020): e1921618.

110 **ONE RECENT ANALYSIS CALCULATED THAT FOR EVERY ONE HUNDRED PEOPLE WITH HIGH LDL CHOLESTEROL:** Markku Laakso and Lilian Fernandes Silva, "Statins and Risk of Type 2 Diabetes: Mechanism and Clinical Implications," *Frontiers in Endocrinology* 14 (September 19, 2023): 1239335.

CHAPTER 5: GUT INSTINCT

114 **JOHN SPETH, AN ANTHROPOLOGIST:** John D. Speth and Eugene Morin, "Putrid Meat in the Tropics: It Wasn't Just for Inuit," *PaleoAnthropology* 2022, no. 2 (October 2022): https://doi.org/10.48738/2022.iss2.114.

117 **PLAQUE STUCK TO THE TEETH OF PALEOLITHIC HUMANS TYPICALLY CONTAINS GRAINS AND STARCH:** Amanda G. Henry, Alison S. Brooks, Dolores R. Piperno, "Plant Foods and the Dietary Ecology of Neanderthals and Early Modern Humans," *Journal of Human Evolution* 69 (April 2014): 44–54.

117 **THE HUMAN STOMACH IS INCREDIBLY ACIDIC, WITH A PH OF JUST 1.5:** DeAnna E. Beasley et al., "The Evolution of Stomach Acidity and Its Relevance to the Human Microbiome," *PLoS One* 10, no. 7 (July 29, 2015): e013 4116.

119 **COPIES OF THE GENE THAT MAKES SALIVARY AMYLASE:** George H. Perry et al., "Diet and the Evolution of Human Amylase Gene Copy Number Variation," *Nature Genetics* 39, no. 10 (October 2007): 1256–60.

119 **"LACTASE PERSISTENCE" EVOLVED:** Tishkoff et al., "Convergent Adaptation of Human Lactase Persistence in Africa and Europe," *Nature Genetics* 39, no. 1 (Janary 2007): 31 – 40.

119 **RECENT STUDY OF OVER THREE HUNDRED THOUSAND ADULTS IN THE UNITED KINGDOM:** Richard P. Evershed et al., "Dairying, Diseases and the Evolution of Lactase Persistence in Europe," *Nature* 608, no. 7922 (August 2022): 336–45.

121 **HUMAN INTESTINAL TRACT IS TWENTY-SEVEN FEET (EIGHT METERS) LONG:** Amanda McGrosky et al., "Gross Intestinal Morphometry and Allometry in Primates," *American Journal of Primatology* 81, no. 8 (August 2019): e23035.

121 **NEARLY TWICE THE CALORIES PER GRAM OF A TYPICAL PRIMATE DIET:** Bruno Simmen et al., "Primate Energy Input and the Evolutionary Transition to Energy-Dense Diets in Humans," *Proceedings of the Royal Society B: Biological Sciences* 284, no. 1856 (June 14, 2017): 20170577.

124 **INCLUDING OUR ORANGUTAN COUSINS:** Cheryl D. Knott, "Changes in Orangutan Caloric Intake, Energy Balance, and Ketones in Response to Fluctuating Fruit Availability," *International Journal of Primatology* 19 (December 1998): 1061–79.

124 **LESS THAN FIFTY GRAMS PER DAY:** "Diet Review: Ketogenic Diet for Weight Loss," Harvard T. H. Chan School of Public Health, accessed February 13, 2024, https://www.hsph.harvard.edu/nutritionsource/healthy-weight/diet-reviews/ketogenic-diet/.

124 **INSULIN PRODUCTION IS TRIGGERED BY RISING GLUCOSE LEVELS:** Zhuo Fu, Elizabeth R. Gilbert, and Dongmin Liu, "Regulation of Insulin Synthesis and Secretion and Pancreatic Beta-Cell Dysfunction in Diabetes," *Current Diabetes Reviews* 9, no. 1 (January 1, 2013): 25–53.

126 **TYPICAL AMERICAN ADULT GAINS A POUND OR TWO PER YEAR:** Larry A. Tucker and Kayla Parker, "10-Year Weight Gain in 13,802 US Adults: The Role of Age, Sex, and Race," *Journal of Obesity* 2022 (May 6, 2022): 7652408.

127 **HADZA GET MORE PHYSICAL ACTIVITY IN A DAY THAN MOST AMERICANS GET IN A WEEK:** David A. Raichlen et al., "Physical Activity Patterns and Biomarkers of Cardiovascular Disease Risk in Hunter-Gatherers," *American Journal of Human Biology* 29, no. 2 (March 2017): e22919.

128 **HADZA MEN AND WOMEN IN OUR SAMPLE WERE NO DIFFERENT FROM THOSE OF U.S. ADULTS:** Herman Pontzer et al., "Hunter-Gatherer Energetics and Human Obesity," *PLoS One* 7, no. 7 (July 5, 2012): e40503.

128 **ENERGY EXPENDITURES AMONG WOMEN IN RURAL NIGERIA:** Amy Luke et al., "Energy Expenditure and Adiposity in Nigerian and African-American Women," *Obesity* 16, no. 9 (September 2008): 2148–54.

128 **LARGE META-ANALYSIS OUT OF LUKE'S LAB:** Amy Luke et al., "Energy Expenditure in Adults Living in Developing Compared with Industrialized Countries: A Meta-Analysis of Doubly Labeled Water Studies," *American Journal of Clinical Nutrition* 93, no. 2 (February 2011): 427–41.

128 **TSIMANE, SHUAR, AND DAASANACH, AND POPULATIONS IN INDUSTRIALIZED COMMUNITIES:** Tsimane: Michael D. Gurven et al., "High Resting Metabolic Rate among Amazonian Forager-Horticulturalists Experiencing High Pathogen Burden," *American Journal of Biological Anthropology* 161, no. 3 (November 2016): 414–25; Shuar: Samuel S. Urlacher et al., "Childhood Daily Energy Expenditure Does Not Decrease with Market Integration and Is Not Related to Adiposity in Amazonia," *Journal of Nutrition* 151, no. 3 (March 2021): 695–704; Daasanach: Amanda McGrosky et al., "Total Daily Energy Expenditure and Elevated Water Turnover in a Small-Scale Semi-Nomadic Pastoralist Society from Northern Kenya," *Annals of Human Biology* 51, no. 1 (March 2024): 2310724.

130 **TYPICAL MAN IN AN INDUSTRIALIZED POPULATION LIKE THE U.S.:** Herman Pontzer et al., "Daily Energy Expenditure through the Human Life Course," *Science* 373, no. 6556 (August 13, 2021): 808–12.

131 **HERITABILITY OF METABOLIC RATE:** M. J. Müller and C. Geisler, "From the Past to Future: From Energy Expenditure to Energy Intake to Energy Expenditure," *European Journal of Clinical Nutrition* 71, no. 3 (March 2017): 358–64.

132 **STUDENTS WITH HIGH ANXIETY LEVELS BURNED ABOUT 6 PERCENT MORE ENERGY:** W. D. Schmidt et al., "Resting Metabolic Rate Is Influenced by Anxiety in College Men," *Journal of Applied Physiology* 80, no. 22 (February 1996): 638–42.

132 **VARIATION IN METABOLIC RATE WE SEE AMONG INDIVIDUALS, IT'S STABLE:** Rebecca Rimbach et al., "Total Energy Expenditure Is Repeatable in Adults but Not Associated with Short-Term Changes in Body Composition," *Nature Communications* 13, no. 1 (January 10, 2022): 99.

133 **BLAMES SUGARS AND OTHER CARBOHYDRATES FOR THE OBESITY CRISIS:** David S. Ludwig et al., "The Carbohydrate-Insulin Model: A Physiological

Perspective on the Obesity Pandemic," *American Journal of Clinical Nutrition* 114, no. 6 (December 1, 2021): 1873–85.

134 **HEAD-TO-HEAD TRIALS:** Michael L. Dansinger et al., "Comparison of the Atkins, Ornish, Weight Watchers, and Zone Diets for Weight Loss and Heart Disease Risk Reduction: A Randomized Trial," *JAMA* 293, no. 1 (January 5, 2005): 43–53.

134 **PEOPLE FIND HIGHER-PROTEIN FOODS MORE SATIATING:** S. H. Holt et al., "A Satiety Index of Common Foods," *European Journal of Clinical Nutrition* 49, no. 9 (September 1995): 675–90.

135 **RESEARCHERS HAVE FOCUSED ON ENERGY DENSITY:** R. J. Stubbs and S. Whybrow, "Energy Density, Diet Composition and Palatability: Influences on Overall Food Energy Intake in Humans," *Physiology & Behavior* 81, no. 5 (July 2004): 755–64.

135 **HIS GROUP FED STUDY PARTICIPANTS TWO DIFFERENT MENUS:** Kevin D. Hall et al., "Ultra-Processed Diets Cause Excess Calorie Intake and Weight Gain: An Inpatient Randomized Controlled Trial of Ad Libitum Food Intake," *Cell Metabolism* 30, no. 1 (July 2, 2019): 67–77.e3.

136 **FAMILY STUDIES OF OBESITY REPORT HERITABILITY ESTIMATES:** Thomas J. Hoffmann et al., "A Large Multiethnic Genome-Wide Association Study of Adult Body Mass Index Identifies Novel Loci," *Genetics* 210, no. 2 (October 2018): 499–515.

137 **CUTOFFS FOR ONE POPULATION MIGHT NOT BE APPROPRIATE FOR ANOTHER:** "Ethnic Differences in BMI and Disease Risk," Harvard T. H. Chan School of Public Health, accessed February 13, 2024, https://www.hsph.harvard.edu/obesity-prevention-source/ethnic-differences-in-bmi-and-disease-risk/.

138 **CALLED FOR BMI TO BE ABANDONED AS AN OUTDATED AND RACIST METRIC:** Adele Jackson-Gibson, "The Racist and Problematic History of the Body Mass Index," *Good Housekeeping*, February 23, 2021, https://www.goodhousekeeping.com/health/diet-nutrition/a35047103/bmi-racist-history/.

138 **RECENT STUDY OF OVER 3.6 MILLION ADULTS IN THE UNITED KINGDOM:** Krishnan Bhaskaran et al., "Association of BMI with Overall and Cause-Specific Mortality: A Population-Based Cohort Study of 3·6 Million Adults in the UK," *Lancet Diabetes & Endocrinology* 6, no. 12 (December 2018): 944–53.

139 **ONE RECENT STUDY, MEN AND WOMEN WITH A BMI OVER 30:** Theresia M. Schnurr et al., "Obesity, Unfavourable Lifestyle and Genetic Risk of Type 2 Diabetes: A Case-Cohort Study," *Diabetologia* 63, no. 7 (July 2020): 1324–32.

141 **STRESS CAN PUSH US TO OVEREAT:** Oh-Ryeong Ha and Seung-Lark Lim, "The Role of Emotion in Eating Behavior and Decisions," *Frontiers in Psychology* 14 (December 7, 2023): 1265074.

141 **OVERWEIGHT MEN AND WOMEN WERE COACHED TO IMPROVE THEIR SLEEP HABITS:** Esra Tasali et al., "Effect of Sleep Extension on Objectively Assessed Energy Intake among Adults with Overweight in Real-Life Settings: A Randomized Clinical Trial," *JAMA Internal Medicine* 182, no. 4 (April 1, 2022): 365–74.

141 **ADULTS WITH OBESITY LOSE AN AVERAGE OF 15 TO 20 PERCENT OF THEIR BODY WEIGHT ON THESE DRUGS:** Domenica Rubino et al. for the STEP 4 Investigators, "Effect of Continued Weekly Subcutaneous Semaglutide vs Placebo on Weight Loss Maintenance in Adults with Overweight or Obesity: The STEP 4 Randomized Clinical Trial," *JAMA* 325, no. 14 (April 13, 2021): 1414–25.

143 **HEALTHIEST DIET HAS NOTHING TO DO WITH YOUR BLOOD TYPE:** Leila Cusack et al., "Blood Type Diets Lack Supporting Evidence: A Systematic Review," *American Journal of Clinical Nutrition* 98, no. 1 (July 2013): 99–104.

143 **EVIDENCE FOR PERSONALIZED DIETS BASED ON YOUR DNA IS EQUALLY THIN:** Christoph Höchsmann et al., "The Personalized Nutrition Study (POINTS): Evaluation of a Genetically Informed Weight Loss Approach, a Randomized Clinical Trial," *Nature Communications* 14, no. 1 (October 9, 2023): 6321.

CHAPTER 6: MUSCLE AND BONE

154 **WORLD-CLASS DECATHLETES:** Raoul Van Damme et al., "Performance Constraints in Decathletes," *Nature* 415, no. 6873 (February 14, 2002): 755–56.

154 **TRAINING ALSO SHAPES THE BRAIN:** David C. Hughes, Stia Ellefsen, and Keith Baar, "Adaptations to Endurance and Strength Training," *Cold Spring Harbor Perspectives in Medicine* 8, no. 6 (June 1, 2018): a029769.

155 **PEOPLE WHO EXERCISE HAVE LESS INFLAMMATION:** Herman Pontzer, "Energy Constraint as a Novel Mechanism Linking Exercise and Health," *Physiology* 33, no. 6 (November 1, 2018): 384–93.

156 **EXERCISE PROMOTES BRAIN HEALTH:** David A. Raichlen and Gene E. Alexander, "Adaptive Capacity: An Evolutionary Neuroscience Model Linking Exercise, Cognition, and Brain Health," *Trends in Neurosciences* 40, no. 7 (July 2017): 408–21.

156 **HADZA, PEOPLE TYPICALLY SQUAT OR KNEEL:** David A. Raichlen et al., "Sitting, Squatting, and the Evolutionary Biology of Human Inactivity," *Proceedings of the National Academy of Sciences* 117, no. 13 (March 31, 2020): 7115–21.

156 **EXERCISE IMPACTS THE PRODUCTION OF THYROID HORMONE:** Christopher L. Klasson, Srishti Sadhir, and Herman Pontzer, "Daily Physical Activity Is

Negatively Associated with Thyroid Hormone Levels, Inflammation, and Immune System Markers among Men and Women in the NHANES Dataset," *PLoS One* 17, no. 7 (July 6, 2022): e0270221.

157 **"RELATIVE ENERGY DEFICIENCY IN SPORT," OR RED-S:** Trent Stellingwerff et al., "Overtraining Syndrome (OTS) and Relative Energy Deficiency in Sport (RED-S): Shared Pathways, Symptoms and Complexities," *Sports Medicine* 51, no. 11 (November 2021): 2251–80.

158 **HELPS TO KEEP YOU MENTALLY SHARP AS YOU AGE:** Brian Duy Ho et al., "Associations between Physical Exercise Type, Fluid Intelligence, Executive Function, and Processing Speed in the Oldest-Old (85 +)," *GeroScience* 46, no. 1 (February 2024): 491–503.

159 **DAN LIEBERMAN, AN EVOLUTIONARY ANTHROPOLOGIST:** Dennis M. Bramble and Daniel E. Lieberman, "Endurance Running and the Evolution of *Homo*," *Nature* 432, no. 7015 (November 18, 2004): 345–52.

159 **BEATING ALL BUT THIRTEEN OF HIS EQUINE COMPETITORS:** Catherine de Lange, "Humans Are Better Endurance Runners than Any Other Animal," *New Scientist*, December 7, 2016.

159 **NICK COURY WON THE WHOLE THING:** "Man against Horse," *Radiolab*, December 28, 2019, https://radiolab.org/podcast/man-against-horse.

160 **MORE THAN HALF OF THE MUSCLE IN THEIR LEGS IS FAST-TWITCH:** Samantha R. Queeno et al., "Human and African Ape Myosin Heavy Chain Content and the Evolution of Hominin Skeletal Muscle," *Comparative Biochemistry and Physiology, Part A: Molecular & Integrative Physiology* 281 (July 2023): 111415.

160 **MORE THAN TRIPLED THE MILEAGE COVERED EACH DAY:** Herman Pontzer, "Locomotor Ecology and Evolution in Chimpanzees and Humans," in *Chimpanzees and Human Evolution*, ed. Martin N. Muller, Richard W. Wrangham, and David R. Pilbeam (Cambridge, MA: Harvard University Press, 2017), 259–85.

161 **TOP MALE ATHLETES PERFORM ABOUT 10 TO 30 PERCENT BETTER:** Sandra K. Hunter et al., "The Biological Basis of Sex Differences in Athletic Performance: Consensus Statement for the American College of Sports Medicine," *Medicine & Science in Sports & Exercise* 55, no. 12 (December 1, 2023): 2328–60.

162 **STUDIES INVESTIGATING THE IMPACT OF TESTOSTERONE ON MUSCLE:** David J. Handelsman, Angelica L. Hirschberg, and Stephane Bermon, "Circulating Testosterone as the Hormonal Basis of Sex Differences in Athletic Performance," *Endocrine Reviews* 39, no. 5 (October 1, 2018): 803–29.

162 **STEROIDS, THE SCOURGE OF COMPETITIVE SPORTS:** "Anabolic Steroid," Wikipedia, accessed February 1, 2024, https://en.wikipedia.org/wiki/Anabolic_steroid.

163 ***ACTN3*, A GENE:** Craig Pickering and John Kiely, "ACTN3: More than Just a Gene for Speed," *Frontiers in Physiology* 8 (December 18, 2017): 1080.

163 **ANOTHER GENE ASSOCIATED WITH ATHLETIC PERFORMANCE IS *ACE*:** Alun Jones, Hugh E. Montgomery, and David R. Woods, "Human Performance: A Role for the ACE Genotype?," *Exercise and Sport Sciences Reviews* 30, no. 4 (October 2002): 184–90.

164 **STEADILY INCREASED THE NUMBER OF GENES IMPLICATED IN STRENGTH AND ENDURANCE:** Celal Bulgay et al., "Exome-Wide Association Study of Competitive Performance in Elite Athletes," *Genes* 14, no. 3 (March 6, 2023): 660.

164 **ALLELES EVEN AFFECT YOUR *RESPONSE* TO EXERCISE:** Camilla J. Williams et al., "Genes to Predict VO_2max Trainability: A Systematic Review," *BMC Genomics* 18, suppl. 8 (November 14, 2017): 831.

164 **STUDIES HAVE IDENTIFIED ALLELES ASSOCIATED WITH HOW MUCH PEOPLE EXERCISE:** Lene Aasdahl et al., "Genetic Variants Related to Physical Activity or Sedentary Behaviour: A Systematic Review," *International Journal of Behavioral Nutrition and Physical Activity* 18, no. 1 (January 22, 2021): 15.

165 **SEVERAL ATHLETES HAD GOTTEN TANTALIZINGLY CLOSE:** "Mile Run," Wikipedia, accessed February 1, 2024, https://en.wikipedia.org/wiki/Mile_run.

166 **ROGER BANNISTER:** "Roger Bannister," Wikipedia, accessed February 1, 2024, https://en.wikipedia.org/wiki/Roger_Bannister.

168 **THE SCIENCE OF EXCELLENCE IN A FOUNDATIONAL PAPER:** K. Anders Ericsson, Ralph T. Krampe, Clemens Tesch-Römer, "The Role of Deliberate Practice in the Acquisition of Expert Performance," *Psychological Review* 100, no. 3 (July 1993): 363–406.

168 **DARTS, ARCHERY, SWIMMING, BOWLING:** Ben Blatt, "Studying the Limits of Human Perfection, through Darts," *New York Times*, August 5, 2023.

169 **JOE BEIMEL:** "Joe Beimel," Baseball Reference, accessed February 13, 2024, https://www.baseball-reference.com/players/b/beimejo01.shtml.

170 **GENETIC INFLUENCE ON THE ABILITY TO KEEP RHYTHM:** Marla Niarchou et al., "Genome-Wide Association Study of Musical Beat Synchronization Demonstrates High Polygenicity," *Nature Human Behaviour* 6, no. 9 (September 2022): 1292–1309.

171 **25 PERCENT OF MAJOR LEAGUE PITCHERS ARE LEFT-HANDED:** Josh Levitt, "Baseball Analysis: Why Lefties Rule," Bleacher Report, July 2, 2009, https://bleacherreport.com/articles/210701-lefties-rule.

171 **HERITABILITY IS PRETTY LOW:** Gabriel Cuellar-Partida et al., "Genome-Wide Association Study Identifies 48 Common Genetic Variants Associated with Handedness," *Nature Human Behaviour* 5, no. 1 (January 2021): 59–70.

171 **THE BEST ATHLETES TEND TO HAVE BIRTHDAYS EARLY IN THE YEAR:** Paolo Riccardo Brustio et al., "Relative Age Influences Performance of World-Class Track and Field Athletes Even in the Adulthood," *Frontiers in Psychology* 10 (June 18, 2019): 1395.

174 **COACHES RESISTED THE IDEA THAT BLACK FOOTBALL PLAYERS COULD BE**

GREAT QUARTERBACKS: Bobby Gehlen and Ivan Pereira, "The Decadeslong Struggle to Break NFL's Quarterback Color Barrier Is Subject of New Book," ABC News, October 10, 2022, https://abcnews.go.com/Sports/decadeslong-struggle-break-nfls-quarterback-color-barrier-subject/story?id=91167615#.

CHAPTER 7: TURNING ENERGY INTO OFFSPRING

178 **WIFE'S MENSTRUAL BLOOD WILL RUIN HER HUSBAND'S ARROW POISON:** Camilla Power, "Hadza Gender Rituals—Epeme and Maitoko—Considered as Counterparts," *Hunter Gatherer Research* 1, no. 3 (September 2015): 333–58.

180 **ACKNOWLEDGE AND EVEN REVERE TRANSGENDER OR THIRD GENDER INDIVIDUALS:** "A Map of Gender-Diverse Cultures," Independent Lens, accessed February 13, 2024, https://www.pbs.org/independentlens/content/two-spirits_map-html/.

184 **THESE CLEAR DISTINCTIONS CAN GET COMPLICATED:** David E. Sandberg and Melissa Gardner, "Differences/Disorders of Sex Development: Medical Conditions at the Intersection of Sex and Gender," *Annual Review of Clinical Psychology* 18 (May 9, 2022): 201–31.

185 **"CONGENITAL ADRENAL HYPERPLASIA":** Ş. Savaş-Erdeve et al., "Clinical Characteristics of 46 XX Males with Congenital Adrenal Hyperplasia," *Journal of Clinical Research in Pediatric Endocrinology* 13, no. 2 (June 2, 2021): 180–86.

185 **THEY AFFECT APPROXIMATELY ONE IN FIVE THOUSAND NEWBORNS EACH YEAR:** Mary García-Acero et al., "Disorders of Sexual Development: Current Status and Progress in the Diagnostic Approach," *Current Urology* 13, no. 4 (January 2020): 169–78.

186 **THREE HORMONE TSUNAMIS:** Marianne Becker and Volker Hesse, "Minipuberty: Why Does It Happen?," *Hormone Research in Paediatrics* 93, no. 2 (June 2020): 76–84.

187 **BRAINS IN BOTH MALES AND FEMALES ARE AFFECTED BY THE FLOOD OF HORMONES:** Erika E. Forbes and Ronald E. Dahl, "Pubertal Development and Behavior: Hormonal Activation of Social and Motivational Tendencies," *Brain and Cognition* 72, no. 1 (February 2010): 66–72.

190 **OVER 150 MILLION TIMES IN HUMANS AROUND THE WORLD EACH YEAR:** There are 130 to 140 million births globally each year (Our World in Data, accessed February 14, 2024, https://ourworldindata.org/births-and-deaths), and many additional pregnancies that end in miscarriage or abortion.

192 **STRESS IN OUR MODERN WORLD CAN NEGATIVELY IMPACT FERTILITY:** Amelia K. Wesselink et al., "Perceived Stress and Fecundability: A Preconception

Cohort Study of North American Couples," *American Journal of Epidemiology* 187, no. 12 (December 1, 2018): 2662–71.

193 **40 PERCENT OF MALES IN THEIR FORTIES EXPERIENCE ERECTILE DYSFUNCTION:** Monica G. Ferrini, Nestor F. Gonzalez-Cadavid, and Jacob Rajfer, "Aging Related Erectile Dysfunction—Potential Mechanism to Halt or Delay Its Onset," *Translational Andrology and Urology* 6, no. 1 (February 2017): 20–27.

193 **ACTOR AL PACINO RECENTLY FATHERED A CHILD AT THE AGE OF EIGHTY-THREE:** "List of Oldest Fathers," Wikipedia, accessed February 1, 2024, https://en.wikipedia.org/wiki/List_of_oldest_fathers.

194 **HUMAN MALES ARE PRETTY UNIMPRESSIVE BY COMPARISON:** J. Michael Plavcan, "Sexual Size Dimorphism, Canine Dimorphism, and Male-Male Competition in Primates: Where Do Humans Fit In?," *Human Nature* 23, no. 1 (March 2012): 45–67.

195 **THESE DIFFERENCES MAINLY EMERGE DURING PUBERTY AND ARE CLEARLY RELATED TO THE EFFECTS OF TESTOSTERONE:** David J. Handelsman, Angelica L. Hirschberg, and Stephane Bermon, "Circulating Testosterone as the Hormonal Basis of Sex Differences in Athletic Performance," *Endocrine Reviews* 39, no. 5 (October 1, 2018): 803–29.

195 **DIFFERS BY AROUND 10 PERCENT AS WELL, AND IS OVER 20 PERCENT IN SOME SPORTS:** Sandra K. Hunter et al., "The Biological Basis of Sex Differences in Athletic Performance: Consensus Statement for the American College of Sports Medicine," *Medicine & Science in Sports & Exercise* 55, no. 12 (December 1, 2023): 2328–60.

196 **AFTER TWELVE MONTHS OF TREATMENT THE REDUCTION IS USUALLY AROUND 5 PERCENT IN NONATHLETES:** Emma N. Hilton and Tommy R. Lundberg, "Transgender Women in the Female Category of Sport: Perspectives on Testosterone Suppression and Performance Advantage," *Sports Medicine* 51, no. 2 (February 2021): 199–214.

196 **SINCE THEY'RE MORE LIKELY TO PARTICIPATE IN SPORTS:** Robert O. Deaner et al., "A Sex Difference in the Predisposition for Physical Competition: Males Play Sports Much More than Females Even in the Contemporary U.S.," *PLoS One* 7, no. 11 (November 14, 2012): e49168.

197 **THE DETAILED ANATOMY OF MALE AND FEMALE BRAINS:** Siyuan Liu et al., "Integrative Structural, Functional, and Transcriptomic Analyses of Sex-Biased Brain Organization in Humans," *Proceedings of the National Academy of Sciences* 117, no. 31 (August 4, 2020): 18788–98.

197 **SOME PATTERNS DO EMERGE:** Ethan Zell, Zlatan Krizan, and Sabrina R. Teeter, "Evaluating Gender Similarities and Differences Using Metasynthesis," *American Psychologist* 70, no. 1 (January 2015): 10–20.

198 **THE GENDER OF PINK:** Jeanne Maglaty, "When Did Girls Start Wearing Pink?," *Smithsonian*, April 7, 2011.

199 **MORE THAN 99 PERCENT OF ADULTS IDENTIFY AS THE GENDER ALIGNED WITH THEIR SEX:** Jody L. Herman, Andrew R. Flores, and Kathryn K. O'Neill, "How Many Adults and Youth Identify as Transgender in the United States?," Williams Institute, UCLA School of Law, June 2022, https://williamsinstitute.law.ucla.edu/publications/trans-adults-united-states/.

199 **ADULTS WHO IDENTIFY AS EITHER GAY OR LESBIAN AT ABOUT 4 PERCENT:** Andrew R. Flores and Kerith J. Conron, "Adult LGBT Population in the United States," Williams Institute, UCLA School of Law, December 2023, https://williamsinstitute.law.ucla.edu/publications/adult-lgbt-pop-us/.

199 **A SCHOOL OF THOUGHT IN PSYCHOLOGY AND SOCIOLOGY:** Mari Mikkola, "Feminist Perspectives on Sex and Gender," in *The Stanford Encyclopedia of Philosophy*, ed. Edward N. Zalta and Uri Nodelman, Fall 2023, https://plato.stanford.edu/archives/fall2023/entries/feminism-gender/.

201 **MALES (XY) WITH NORMAL TESTOSTERONE PRODUCTION AND RECEPTORS DO NOT FORM A PENIS:** Heino F. L. Meyer-Bahlburg, "Gender Identity Outcome in Female-Raised 46,XY Persons with Penile Agenesis, Cloacal Exstrophy of the Bladder, or Penile Ablation," *Archives of Sexual Behavior* 34, no. 4 (August 2005): 423–38.

202 **GUIDANCE IN THESE CASES HAS SHIFTED AWAY FROM SURGICAL INTERVENTION AND FEMALE GENDER ASSIGNMENT:** "Aphallia," Cleveland Clinic, accessed February 14, 2024, https://my.clevelandclinic.org/health/diseases/24246-aphallia.

202 **5-ARD PROVIDE ANOTHER EXAMPLE OF HORMONAL INFLUENCE ON GENDER IDENTITY:** Ramesh Babu and Utsav Shah, "Gender Identity Disorder (GID) in Adolescents and Adults with Differences of Sex Development (DSD): A Systematic Review and Meta-Analysis," *Journal of Pediatric Urology* 17, no. 1 (February 2021): 39–47.

202 **WORK BY DAVID PUTS:** Ashlyn Swift-Gallant et al., "Organizational Effects of Gonadal Hormones on Human Sexual Orientation," *Adaptive Human Behavior and Physiology* 9, no. 4 (September 2023): 344–70.

202 **FEMALES EXPOSED TO HIGH LEVELS OF ANDROGENS DUE TO CONGENITAL ADRENAL HYPERPLASIA:** Sheri A. Berenbaum and Adriene M. Beltz, "Sexual Differentiation of Human Behavior: Effects of Prenatal and Pubertal Organizational Hormones," *Frontiers in Neuroendocrinology* 32, no. 2 (April 2011): 183–200.

202 **SOME EVIDENCE THAT THEY ARE MORE LIKELY TO IDENTIFY AS BOYS:** Vicki Pasterski et al., "Increased Cross-Gender Identification Independent of Gender Role Behavior in Girls with Congenital Adrenal Hyperplasia: Results from a Standardized Assessment of 4- to 11-Year-Old Children," *Archives of Sexual Behavior* 44, no. 5 (July 2015): 1363–75.

203 **TRANSGENDER IDENTITY AND SAME-SEX ATTRACTION ARE MODESTLY HERITABLE:** Ferdinand J. O. Boucher and Tudor I. Chinnah, "Gender Dysphoria: A Review Investigating the Relationship between Genetic Influences and

Brain Development," *Adolescent Health, Medicine and Therapeutics* 11 (August 2020): 89–99.

203 **ALLELES THAT ARE WEAKLY ASSOCIATED WITH SAME-SEX ATTRACTION:** Andrea Ganna et al., "Large-Scale GWAS Reveals Insights into the Genetic Architecture of Same-Sex Sexual Behavior," *Science* 365, no. 6456 (August 2019): eaat7693.

206 **BIOLOGIST ANNE FAUSTO-STERLING:** "Gender & Sexuality," Anne Fausto-Sterling, accessed February 1, 2024, https://www.annefaustosterling.com/fields-of-inquiry/gender/.

206 **BIOLOGICAL ANTHROPOLOGIST AGUSTÍN FUENTES:** Agustín Fuentes, "Here's Why Human Sex Is Not Binary," *Scientific American*, May 1, 2023, https://www.scientificamerican.com/article/heres-why-human-sex-is-not-binary/.

207 ***FAFINE* IN SAMOA, FOR EXAMPLE:** "Fa'afafine," Wikipedia, accessed February 1, 2024, https://en.wikipedia.org/wiki/Fa'afafine.

209 **SLOWER THAN THE OLYMPIC-QUALIFYING TIMES FOR MEN:** Men's qualifying times: Gordon Mack, "2020 Olympic Qualifying Standards Released," FloTrack, March 10, 2019, https://www.flotrack.org/articles/6394026-2020-olympic-qualifying-standards-released. Women's track records: "Women's World Records," *Track & Field News*, accessed February 14, 2024, https://trackandfieldnews.com/records/womens-world-records/.

209 **THE CURRENT FEMALE RECORD, 4:07.9:** "Mile Run," Wikipedia, accessed February 1, 2024, https://en.wikipedia.org/wiki/Mile_run.

209 **STUDY BY WORLD RUGBY:** "Transgender Guidelines," World Rugby, October 9, 2020, https://www.world.rugby/the-game/player-welfare/guidelines/transgender.

209 **OF ARCHERY AND OTHER SHOOTING SPORTS:** Blair R. Hamilton et al., "Integrating Transwomen Athletes into Elite Competition: The Case of Elite Archery and Shooting," *European Journal of Sport Science* 21, no. 11 (November 2021): 1500–9.

209 **THERE COULD BE A *FEMALE* ADVANTAGE IN ULTRAMARATHONS:** Ned Rozell, "Women May Have Advantage in the Long Run," UAF News and Information, March 10, 2022, https://www.uaf.edu/news/women-may-have-advantage-in-the-long-run.php#.

209 **CASTER SEMENYA:** "Caster Semenya," Wikipedia, accessed February 1, 2024, https://en.wikipedia.org/wiki/Caster_Semenya.

211 **HOW HANDEDNESS DEVELOPS:** Carolien de Kovel, Amaia Carrión-Castillo, and Clyde Francks, "A Large-Scale Population Study of Early Life Factors Influencing Left-Handedness," *Scientific Reports* 9, no. 1 (January 24, 2019): 584.

211 **THE OLD STORY OF BEING LEFT-HANDED HELD THAT IT WAS DEEPLY UNNATURAL:** Howard I. Kushner, "Why Are There (Almost) No Left-Handers in China?," *Endeavour* 37, no. 2 (June 2013): 71–81.

212 **FIGURE 7.4:** "2020 United States Presidential Election," Wikipedia, accessed February 1, 2024, https://en.wikipedia.org/wiki/2020_United_States_presidential_election; "What Percentage of the Population Is Transgender in 2024?," World Population Review, accessed February 1, 2024, https://worldpopulationreview.com/state-rankings/transgender-population-by-state; Flores and Conron, "Adult LGBT Population in the United States."

213 **ATTITUDES AROUND HOMOSEXUALITY SHIFTED:** "A Survey of LGBT Americans," Pew Research Center, June 13, 2013, https://www.pewresearch.org/social-trends/2013/06/13/a-survey-of-lgbt-americans/.

213 **2019 STUDY OF SEXUAL ORIENTATION USING THE UK BIOBANK:** Ganna et al., "Large-Scale GWAS Reveals Insights into the Genetic Architecture of Same-Sex Sexual Behavior."

CHAPTER 8: ENVIRONMENTAL PROTECTION

216 **EXCESSIVE HEAT KILLS TENS OF THOUSANDS:** Noah Scovronick et al., "The Association between Ambient Temperature and Mortality in South Africa: A Time-Series Analysis," *Environmental Research* 161 (February 2018): 229–35.

217 **THESE WERE THE EARLIEST MAMMALS:** "Evolution of Mammals," Wikipedia, accessed February 1, 2024, https://en.wikipedia.org/wiki/Evolution_of_mammals.

218 **REPTILE BODY TEMPERATURES CAN FLUCTUATE BY TWENTY DEGREES:** Bayard H. Brattstrom, "Body Temperatures of Reptiles," *American Midland Naturalist* 73, no. 2 (April 1965): 376–422.

218 **PEOPLE VARY BY A DEGREE OR TWO IN THEIR PERSONAL SETTING:** Ziad Obermeyer, Jasmeet K. Samra, and Sendhil Mullainathan, "Individual Differences in Normal Body Temperature: Longitudinal Big Data Analysis of Patient Records," *BMJ* 359 (December 2017): j5468.

219 **HUMAN BODY TEMPERATURE HAS FALLEN OVER THE PAST CENTURY:** Myroslava Protsiv et al., "Decreasing Human Body Temperature in the United States since the Industrial Revolution," *eLife* 9 (January 7, 2020): e49555.

219 **THE SMALL DECLINE IN BASAL METABOLIC RATE:** John R. Speakman et al., "Total Daily Energy Expenditure Has Declined over the Past Three Decades Due to Declining Basal Expenditure, Not Reduced Activity Expenditure," *Nature Metabolism* 5, no. 4 (April 2023): 579–88.

221 **XUEYING ZHANG RECENTLY LED A LARGE COLLABORATIVE RESEARCH STUDY:** Xueying Zhang et al., "Human Total, Basal and Activity Energy Expenditures Are Independent of Ambient Environmental Temperature," *iScience* 25, no. 8 (June 28, 2022): 104682.

221 **PASTORALISTS LIVING WITH THEIR HERDS IN TUVA, SIBERIA:** Adam J. Sellers et al., "High Daily Energy Expenditure of Tuvan Nomadic Pastoralists Living in an Extreme Cold Environment," *Scientific Reports* 12, no. 1 (November 22, 2022): 20127.

221 **TUVAN ADULTS ALSO APPEAR TO VASOCONSTRICT *LESS* THAN OTHER POPULATIONS:** Adam J. Sellers, Dolaana Khovalyg, Wouter van Marken Lichtenbelt, "Thermoregulation of Tuvan Pastoralists and Western Europeans during Cold Exposure," *American Journal of Human Biology* 35, no. 10 (October 2023): e23933.

222 **NEARLY ONE THOUSAND PEOPLE IN THE U.S. DIE EACH YEAR FROM HYPOTHERMIA:** "Climate Change Indicators: Cold-Related Deaths," EPA, accessed February 1, 2024, https://www.epa.gov/climate-indicators/climate-change-indicators-cold-related-deaths.

222 **GIRL IN UTAH WAS PLUCKED FROM AN ICY CREEK:** R. G. Bolte et al., "The Use of Extracorporeal Rewarming in a Child Submerged for 66 Minutes," *JAMA* 260, no. 3 (July 15, 1988): 377–79.

222 **OLDEST-KNOWN COLD-WATER NEAR-DROWNING SURVIVOR:** A. H. Chochinov et al., "Recovery of a 62-Year-Old Man from Prolonged Cold Water Submersion," *Annals of Emergency Medicine* 31, no. 1 (January 1998): 127–31.

222 **105 DEGREES FAHRENHEIT (41°C) IF YOU'RE EXERCISING IN THE HEAT:** Polly Aylwin et al., "Thermoregulatory Responses during Road Races in Hot-Humid Conditions at the 2019 Athletics World Championships," *Journal of Applied Physiology* 134, no. 5 (May 1, 2023): 1300–11.

223 **HEATSTROKE:** "Heatstroke," Mayo Clinic, accessed February 14, 2024, https://www.mayoclinic.org/diseases-conditions/heat-stroke/symptoms-causes/syc-20353581.

223 **HUMAN BODY CAN'T COPE WITH AIR TEMPERATURES ABOVE NINETY-SEVEN DEGREES FAHRENHEIT (36°C):** Daniel J. Vecellio et al., "Evaluating the 35°C Wet-Bulb Temperature Adaptability Threshold for Young, Healthy Subjects (PSU HEAT Project)," *Journal of Applied Physiology* 132, no. 2 (February 1, 2022): 340–45.

223 **HEAT WAVES REGULARLY KILL TENS OF THOUSANDS OF PEOPLE ACROSS THE U.S. AND EUROPE:** U.S.: S. A. M. Khatana, R. M. Werner, and P. W. Groeneveld, "Association of Extreme Heat with All-Cause Mortality in the Contiguous US, 2008–2017, *JAMA Network Open* 5, no. 5 (May 19, 2022): e2212957. Europe: P. Masselot et al., "Excess Mortality Attributed to Heat and Cold: A Health Impact Assessment Study in 854 Cities in Europe," *Lancet Planetary Health* 7, no. 4 (April 2023): e271–e281.

224 **PROFESSIONAL ATHLETES COMPETING FOR THE PODIUM RUN SLOWER IN THE HEAT:** Matthew R. Ely et al., "Impact of Weather on Marathon-Running Performance," *Medicine & Science in Sports & Exercise* 39, no. 3 (March 2007): 487–93.

224 **MORE SWEAT GLANDS THAN ANY OTHER MAMMAL:** Yana G. Kamberov et al.,

"Comparative Evidence for the Independent Evolution of Hair and Sweat Gland Traits in Primates," *Journal of Human Evolution* 125 (December 2018): 99–105.

224 **YOUR SWEAT GLANDS ARE ALSO CLEVER:** Lisa Klous et al., "Sweat Rate and Sweat Composition during Heat Acclimation," *Journal of Thermal Biology* 93 (October 2020): 102697.

226 **INVESTIGATED APE AND HUMAN WATER USE IN A RECENT STUDY:** Herman Pontzer et al., "Evolution of Water Conservation in Humans," *Current Biology* 31, no. 8 (April 26, 2021): 1804–10.e5.

226 **NOSE SHAPE ACROSS HUMAN GROUPS TODAY REFLECTS VARIATION IN CLIMATE AND HUMIDITY:** Arslan A. Zaidi et al., "Investigating the Case of Human Nose Shape and Climate Adaptation," *PLoS Genetics* 13, no. 3 (March 16, 2017): e1006616.

228 **ASHER ROSINGER, AN ANTHROPOLOGIST:** Asher Y. Rosinger, "Human Evolution Led to an Extreme Thirst for Water," *Scientific American*, July 1, 2021, https://www.scientificamerican.com/article/human-evolution-led-to-an-extreme-thirst-for-water/.

228 **PERENNIAL SOURCE OF WORRY AND WORK:** Hilary J. Bethancourt et al., "The Co-Occurrence of Water Insecurity and Food Insecurity among Daasanach Pastoralists in Northern Kenya," *Public Health Nutrition* 26, no. 3 (August 9, 2022): 1–11.

228 **PROVIDING A WELL LITERALLY SAVES ENOUGH ENERGY TO BUILD MORE BABIES:** Mhairi A. Gibson and Ruth Mace, "An Energy-Saving Development Initiative Increases Birth Rate and Childhood Malnutrition in Rural Ethiopia," *PLoS Medicine* 3, no. 4 (April 2006): e87.

228 **GENETICS RESEARCH IN NEIGHBORING COMMUNITIES, BY AMANDA LEA AND JULIAN AYROLES:** A. J. Lea et al., "Adaptations to Water Stress and Pastoralism in the Turkana of Northwest Kenya," *bioRxiv*, preprint (January 19, 2023): https://doi.org/10.1101/2023.01.17.524066.

229 **2 MILLION PEOPLE IN THE U.S. LACK ACCESS TO SAFE TAP WATER:** Ashir Rosinger, "Nearly 60 Million Americans Don't Drink Their Tap Water, Research Suggests—Here's Why That's a Public Health Problem," Conversation, April 15, 2021, https://theconversation.com/nearly-60-million-americans-dont-drink-their-tap-water-research-suggests-heres-why-thats-a-public-health-problem-158483.

229 **BOTTLED WATER BUSINESS MAKES MORE THAN $300 *BILLION* PER YEAR GLOBALLY:** "Bottled Water Market Size & Share Analysis—Growth Trends & Forecasts (2024–2029)," Mordor Intelligence, accessed February 14, 2024, https://www.mordorintelligence.com/industry-reports/bottled-water-market.

231 **NEPHRONS ARE EXQUISITELY SENSITIVE:** Brian Cuzzo, Sandeep A. Padala, Sarah L. Lappin, "Physiology, Vasopressin" (Treasure Island, FL: StatPearls, 2024).

233 **YOSUKE YAMADA LED A BIG INTERNATIONAL STUDY:** Yosuke Yamada et al., "Variation in Human Water Turnover Associated with Environmental and Lifestyle Factors," *Science* 378, no. 6622 (November 25, 2022): 909–15.

235 **CHRONICALLY DEHYDRATING YOURSELF, EVEN A BIT, CAN CAUSE TROUBLE:** Natalia I. Dmitrieva et al., "Middle-Age High Normal Serum Sodium as a Risk Factor for Accelerated Biological Aging, Chronic Diseases, and Premature Mortality," *eBioMedicine* 87 (January 2023): 104404.

235 **A WOMAN RUNNING THE 2002 BOSTON MARATHON DIED:** Christopher S. D. Almond et al., "Hyponatremia among Runners in the Boston Marathon," *New England Journal of Medicine* 352, no. 15 (April 14, 2005): 1550–56.

236 **FREQUENCIES THAT COULD PASS THROUGH THE ATMOSPHERE AND THEN THROUGH SEVERAL FEET OF WATER:** Patrick J. Butler et al., *Animal Physiology: An Environmental Perspective* (New York: Oxford University Press, 2021).

236 **LENSES IN OUR EYES FILTER OUT UV LIGHT:** R. H. Douglas and G. Jeffery, "The Spectral Transmission of Ocular Media Suggests Ultraviolet Sensitivity Is Widespread among Mammals," *Proceedings of the Royal Society B: Biological Sciences* 281, no. 1780 (February 19, 2014): 20132995.

237 **CLAUDE MONET HAD THE LENS IN HIS RIGHT EYE SURGICALLY REMOVED:** Anna Gruener, "The Effect of Cataracts and Cataract Surgery on Claude Monet," *British Journal of General Practice* 65, no. 634 (May 2015): 254–55.

238 **VITAMIN D'S REACH GOES FAR BEYOND THE SKELETON:** "Vitamin D Fact Sheet for Health Professionals," National Institutes of Health, accessed February 14, 2024, https://ods.od.nih.gov/factsheets/VitaminD-HealthProfessional/.

239 **MELANIN IS ANCIENT:** M. E. McNamara et al., "Decoding the Evolution of Melanin in Vertebrates," *Trends in Ecology & Evolution* 36, no. 5 (May 2021): 430–43.

240 **OVER 160 GENES KNOWN TO AFFECT THE PRODUCTION OF MELANIN:** Vivek K. Bajpai et al., "A Genome-Wide Genetic Screen Uncovers Determinants of Human Pigmentation," *Science* 381, no. 6658 (August 11, 2023): eade6289.

240 **SKIN COLOR TRACKED THE PREVAILING CONDITIONS:** Mark D. Lucock, "The Evolution of Human Skin Pigmentation: A Changing Medley of Vitamins, Genetic Variability, and UV Radiation during Human Expansion," *American Journal of Biological Anthropology* 180, no. 2 (February 2023): 252–71.

241 **FIGURE 8.2:** N. G. Jablonski and G. Chaplin, "The Evolution of Human Skin Coloration," *Journal of Human Evolution* 39, no. 1 (July 2000): 57–106; "Modern Scientific Explanations of Human Biological Variation," Britannica, accessed February 15, 2024, https://www.britannica.com/topic/race-human/Modern-scientific-explanations-of-human-biological-variation#/media/1/488030/52059.

242 **GENES THAT DETERMINE HAIR TEXTURE:** Tina Lasisi et al., "Human Scalp Hair as a Thermoregulatory Adaptation," *Proceedings of the National Academy of Sciences* 120, no. 24 (June 13, 2023): e2301760120.

242 **GEOGRAPHY OF GENETIC VARIANTS BROWSER:** "Geography of Genetic Variants Browser," University of Chicago, accessed February 15, 2024, https://popgen.uchicago.edu/ggv/?data=%221000genomes%22&chr=1&pos=222087833.

243 **MOST AMERICANS WHO IDENTIFY AS BLACK HAVE SOME EUROPEAN ANCESTRY:** Katarzyna Bryc et al., "The Genetic Ancestry of African Americans, Latinos, and European Americans across the United States," *American Journal of Human Genetics* 96, no. 1 (January 8, 2015): 37–53.

244 **ABOUT 15 PERCENT OF ALL NEWBORNS IN THE U.S. ARE MIXED-RACE:** Richard D. Alba, "The Surge of Young Americans from Minority-White Mixed Families & Its Significance for the Future," *Dædalus* 150, no. 2 (Spring 2021): 199–214.

244 **YOU SHARE MORE THAN 99.9 PERCENT OF YOUR DNA WITH EVERY OTHER PERSON ON THE PLANET:** Theresa M. Duello et al., "Race and Genetics versus 'Race' in Genetics: A Systematic Review of the Use of African Ancestry in Genetic Studies," *Evolution, Medicine, and Public Health* 9, no. 1 (June 15, 2021): 232–45.

244 **90 PERCENT OF THE ALLELES (GENE VARIANTS) IN THAT POPULATION WILL BE FOUND IN EVERY OTHER POPULATION ON THE PLANET:** James Kitchens and Graham Coop, "Visualizing Human Genetic Diversity," James Kitchens, May 16, 2023, https://james-kitchens.com/blog/visualizing-human-genetic-diversity.

245 **EQUATIONS THAT DOCTORS USE:** Darshali A. Vyas, Leo G. Eisenstein, and David S. Jones, "Hidden in Plain Sight—Reconsidering the Use of Race Correction in Clinical Algorithms," *New England Journal of Medicine* 383, no. 9 (August 27, 2020): 874–82.

246 **BABIES LEARN TO USE SKIN COLOR TO IDENTIFY FACES:** Jessica Sullivan, Leigh Wilton, and Evan P. Apfelbaum, "Adults Delay Conversations about Race Because They Underestimate Children's Processing of Race," *Journal of Experimental Psychology: General* 150, no. 2 (February 2021): 395–400.

247 **TUTSI VIEWED THEIR HUTU NEIGHBORS AS DARKER AND SHORTER:** "Rwanda Genocide of 1994," Britannica, accessed February 15, 2024, https://www.britannica.com/event/Rwanda-genocide-of-1994.

CHAPTER 9: HOW NOT TO DIE (AND WHY YOU WILL ANYWAY)

253 **SUGARS AND FATS CAN INCREASE THE PRODUCTION OF INFLAMMATORY MOLECULES:** Xiao Ma et al., "Excessive Intake of Sugar: An Accomplice of Inflammation," *Frontiers in Immunology* 13 (August 31, 2022): 988481.

254 **THE HYGIENE HYPOTHESIS:** Megan Scudellari, "News Feature: Cleaning Up the Hygiene Hypothesis," *Proceedings of the National Academy of Sciences* 114, no. 7 (February 14, 2017): 1433–36.

254 **HISTAMINE-DRIVEN REACTIONS TO AIRBORNE ALLERGENS WERE ESSENTIALLY UNKNOWN BEFORE 1900:** Thomas A. E. Platts-Mills, "The Allergy Epidemics: 1870–2010," *Journal of Allergy and Clinical Immunology* 136, no. 1 (July 2015): 3–13.

255 **PEOPLE WITH ANTINUCLEAR BODIES IN THEIR BLOOD:** Helen C. S. Meier et al., "Hygiene Hypothesis Indicators and Prevalence of Antinuclear Antibodies in US Adolescents," *Frontiers in Immunology* 13 (January 28, 2022): 789379.

255 **FORTY-FIVE OUT OF EVERY ONE HUNDRED KIDS BORN IN THE U.S. DIED BEFORE THE AGE OF FIVE:** Aaron O'Neill, "Child Mortality Rate (under Five Years Old) in the United States, from 1800 to 2020," Statista, February 2, 2024, https://www.statista.com/statistics/1041693/united-states-all-time-child-mortality-rate/.

256 **LOUIS PASTEUR, OF "PASTEURIZED MILK" FAME:** "Vaccine Development of Louis Pasteur," Britannica, accessed February 15, 2024, https://www.britannica.com/biography/Louis-Pasteur/Vaccine-development.

257 **PEOPLE IN AFRICA AND ASIA HAD FIGURED OUT THAT EARLY EXPOSURE TO SMALLPOX:** Stefan Riedel, "Edward Jenner and the History of Smallpox and Vaccination," *Baylor University Medical Center Proceedings* 18, no. 1 (January 2005): 21–25.

257 **LEADERS IN BOSTON PUSHED INOCULATION DESPITE PROTESTS:** Matthew Niederhuber, "The Fight over Inoculation during the 1721 Boston Smallpox Epidemic," Science in the News, Harvard University, December 31, 2014, https://sitn.hms.harvard.edu/flash/special-edition-on-infectious-disease/2014/the-fight-over-inoculation-during-the-1721-boston-smallpox-epidemic/.

257 **PRESUPPOSES VIRUSES ARE ALIVE:** Luis P. Villarreal, "Are Viruses Alive?" *Scientific American*, August 8, 2008, https://www.scientificamerican.com/article/are-viruses-alive-2004/.

258 **ANOTHER EIGHTY YEARS TO FIGURE OUT *HOW* VACCINATION WORKS:** Max D. Cooper, "The Early History of B Cells," *Nature Reviews Immunology* 15, no. 3 (March 2015): 191–97.

260 **CONNECTION BETWEEN VACCINES AND AUTISM HAS BEEN THOROUGHLY INVESTIGATED AND COMPLETELY DEBUNKED:** "Autism and Vaccines," Centers for Disease Control and Prevention, accessed February 15, 2024, https://www.cdc.gov/vaccinesafety/concerns/autism.html.

261 **REVIEW EXAMINED OVER FORTY RANDOMIZED CONTROL TRIALS:** Carolina Graña et al., "Efficacy and Safety of COVID-19 Vaccines," *Cochrane Database of Systematic Reviews* 12, no. 12 (December 7, 2022): CD015477.

261 **TEN TIMES GREATER FOR UNVACCINATED PEOPLE:** "Monthly Age-Adjusted Rates of COVID-19-Associated Hospitalization by Vaccination Status in Patients Ages ≥18 Years January 2021–January 2023," Centers for Disease Control and Prevention, accessed February 15, 2024, www.cdc.gov/coronavirus/2019-ncov/covid-data/covid-net/hospitalizations-by-vaccination-status-report.pdf.

262 **COVID VACCINE EFFECTIVENESS AND SAFETY IN CHILDREN:** Vanessa Piechotta et al., "Safety and Effectiveness of Vaccines against COVID-19 in Children Aged 5-11 Years: A Systematic Review and Meta-Analysis," *Lancet Child & Adolescent Health* 7, no. 6 (June 2023): 379–91.

262 **VACCINATION RATES ARE FALLING IN COUNTRIES AROUND THE WORLD:** Michael Eisenstein, "Vaccination Rates Are Falling, and It's Not Just the COVID-19 Vaccine that People Are Refusing," *Nature* 612, no. 7941 (December 2022): S44–S46.

262 **VACCINE HESITANCY IS GROWING AMONG U.S. DOG OWNERS:** Matt Motta, Gabriella Motta, Dominik Stecula, "Sick as a Dog? The Prevalence, Politicization, and Health Policy Consequences of Canine Vaccine Hesitancy (CVH)," *Vaccine* 41, no. 41 (September 22, 2023): 5946–50.

262 **REPUBLICAN COUNTIES ARE LESS LIKELY TO GET VACCINATED AGAINST COVID, AND MORE LIKELY TO DIE:** Daniel Wood and Geoff Brumfiel, "Pro-Trump Counties Now Have Far Higher COVID Death Rates. Misinformation Is to Blame," NPR, updated December 5, 2021, https://www.npr.org/sections/health-shots/2021/12/05/1059828993/data-vaccine-misinformation-trump-counties-covid-death-rate.

262 **DEMOCRATIC AND REPUBLICAN VOTERS IN FLORIDA AND OHIO HAD EQUIVALENT MORTALITY RATES:** Jacob Wallace, Paul Goldsmith-Pinkham, and Jason L. Schwartz, "Excess Death Rates for Republican and Democratic Registered Voters in Florida and Ohio during the COVID-19 Pandemic," *JAMA Internal Medicine* 183, no. 9 (September 1, 2023): 916–23.

263 **DNA ANALYSES CARRIED OUT RECENTLY IN CEMETERIES:** Jennifer Klunk et al., "Evolution of Immune Genes Is Associated with the Black Death," *Nature* 611, no. 7935 (November 2022): 312–19.

264 **UNLIKELY TO SEE MUCH OF AN IMMUNE SYSTEM BOOST FROM MULTIVITAMINS:** "Should I Take a Daily Multivitamin?," Nutrition Source, Harvard T. H. Chan School of Public Health, accessed February 15, 2024, https://www.hsph.harvard.edu/nutritionsource/multivitamin/.

264 **VITAMIN C DOESN'T SEEM TO OFFER ADDITIONAL PROTECTION AGAINST THE COMMON COLD:** Harry Hemilä and Elizabeth Chalker, "Vitamin C for Preventing and Treating the Common Cold," *Cochrane Database of Systematic Reviews* 2013, no. 1 (January 31, 2013): CD000980.

265 **THERE ARE NEARLY SIX HUNDRED THOUSAND PEOPLE OVER ONE HUNDRED YEARS OLD:** Katharina Buchholz, "There Are Now More Than Half a Million People Aged 100 or Older around the World," World Economic

Forum, February 17, 2021, https://www.weforum.org/agenda/2021/02/living-to-one-hundred-life-expectancy/.

266 **IN 1900, INFECTIOUS DISEASES WERE THE LEADING KILLERS IN THE U.S.:** David S. Jones, Scott H. Podolsky, Jeremy A. Greene, "The Burden of Disease and the Changing Task of Medicine," *New England Journal of Medicine* 366, no. 25 (June 21, 2012): 2333–38.

266 **REDUCE YOUR RISK FOR SOME CANCERS WITH SIMPLE LIFESTYLE CHANGES:** "Preventing Cancer: Identifying Risk Factors," AACR Cancer Progress Report 2022, American Association for Cancer Research, accessed February 1, 2024, https://cancerprogressreport.aacr.org/progress/cpr22-contents/cpr22-preventing-cancer-identifying-risk-factors/.

267 **CANCER IS MORE LIKELY TO OCCUR IN YOUR COLON:** Cristian Tomasetti and Bert Vogelstein, "Cancer Etiology. Variation in Cancer Risk among Tissues Can Be Explained by the Number of Stem Cell Divisions," *Science* 347, no. 6217 (January 2, 2015): 78–81.

268 **HOURS SPENT SEDENTARY ON THE COUCH OR AT A DESK INCREASE DEMENTIA RISK:** David A. Raichlen et al., "Leisure-Time Sedentary Behaviors Are Differentially Associated with All-Cause Dementia Regardless of Engagement in Physical Activity," *Proceedings of the National Academy of Sciences* 119, no. 35 (August 30, 2022): e2206931119.

268 **DIET STUDIES HAVE BEEN LESS COMPELLING:** "What Do We Know about Diet and Prevention of Alzheimer's Disease?," National Institute on Aging, accessed February 15, 2024, https://www.nia.nih.gov/health/alzheimers-and-dementia/what-do-we-know-about-diet-and-prevention-alzheimers-disease.

268 **SLEEP IS ALSO EMERGING AS AN IMPORTANT LIFESTYLE FACTOR FOR DEMENTIA RISK:** Le Shi et al., "Sleep Disturbances Increase the Risk of Dementia: A Systematic Review and Meta-Analysis," *Sleep Medicine Reviews* 40 (August 2018): 4–16.

269 **FOR ALZHEIMER'S DISEASE, THE ALLELES YOU CARRY AT MORE THAN EIGHTY GENES AFFECT YOUR CHANCES:** "Alzheimer's Disease Genetics Fact Sheet," National Institute on Aging, accessed February 15, 2024, https://www.nia.nih.gov/health/genetics-and-family-history/alzheimers-disease-genetics-fact-sheet.

269 **REMARKABLY LOW PREVALENCE OF ALZHEIMER'S AND OTHER COGNITIVE IMPAIRMENT:** M. Gatz et al., "Prevalence of Dementia and Mild Cognitive Impairment in Indigenous Bolivian Forager-Horticulturalists," *Alzheimer's & Dementia* 19, no. 1 (January 2023): 44–55.

269 ***APOEε4* IS JUST AS COMMON IN THEIR POPULATION AS IT IS IN THE U.S.:** B. C. Trumble et al., "*Apolipoprotein-ε4* Is Associated with Higher Fecundity in a Natural Fertility Population," *Science Advances* 9, no. 32 (August 9, 2023): eade9797.

269 **HYGIENE HYPOTHESIS FOR ALZHEIMER'S DISEASE:** Molly Fox et al., "Hygiene

and the World Distribution of Alzheimer's Disease: Epidemiological Evidence for a Relationship between Microbial Environment and Age-Adjusted Disease Burden," *Evolution, Medicine, and Public Health* 2013, no. 1 (January 2013): 173–86.

270 **ENERGY THE BODY DEVOTES TO MAINTENANCE AND REPAIR VERSUS GROWTH AND REPRODUCTION:** Herman Pontzer and Amanda McGrosky, "Balancing Growth, Reproduction, Maintenance, and Activity in Evolved Energy Economies," *Current Biology* 32, no. 12 (June 20, 2022): R709–R719.

270 **OLDEST PERSON ON RECORD, JEANNE CALMENT:** "Jeanne Calment," Wikipedia, accessed February 1, 2024, https://en.wikipedia.org/wiki/Jeanne_Calment.

271 **MEN TEND TO AGE FASTER AND DIE YOUNGER THAN WOMEN:** Ben A. D. Lendrem et al., "The Darwin Awards: Sex Differences in Idiotic Behaviour," *BMJ* 349 (December 11, 2014): g7094.

271 **LONGEVITY IS HERITABLE:** J. Graham Ruby et al., "Estimates of the Heritability of Human Longevity Are Substantially Inflated due to Assortative Mating," *Genetics* 210, no. 3 (November 2018): 1109–24.

271 **EVOLUTION HAS GIVEN US GRANDMOTHERS:** K. Hawkes et al., "Grandmothering, Menopause, and the Evolution of Human Life Histories," *Proceedings of the National Academy of Sciences* 95, no. 3 (February 3, 1998): 1336–39.

272 **HORMONE REPLACEMENT THERAPY INCREASES THE LIKELIHOOD:** Jane Marjoribanks et al., "Long-Term Hormone Therapy for Perimenopausal and Postmenopausal Women," *Cochrane Database of Systematic Reviews* 1, no. 1 (January 17, 2017): CD004143.

272 **CALORIE RESTRICTION IS AN EFFECTIVE WAY TO EXTEND LIFESPANS:** John R. Speakman and Sharon E. Mitchell, "Caloric Restriction," *Molecular Aspects of Medicine* 32, no. 3 (June 2011): 159–221.

272 **THE SIGNALING MOLECULE AMPK:** Asier González et al., "AMPK and TOR: The Yin and Yang of Cellular Nutrient Sensing and Growth Control," *Cell Metabolism* 31, no. 3 (March 3, 2020): 472–92.

273 **PEOPLE LOSE MORE WEIGHT WHEN ASSIGNED TO TIME-RESTRICTED EATING:** Tanja Črešnovar et al., "Effectiveness of Time-Restricted Eating with Caloric Restriction vs. Caloric Restriction for Weight Loss and Health: Meta-Analysis," *Nutrients* 15, no. 23 (November 24, 2023): 4911.

273 **METFORMIN EXTENDS LIFESPANS IN MICE:** Alejandro Martin-Montalvo et al., "Metformin Improves Healthspan and Lifespan in Mice," *Nature Communications* 4 (January 1, 2013): 2192.

273 **RESVERATROL DOESN'T SEEM TO HAVE ANY EFFECT ON LONGEVITY IN MICE:** Richard A. Miller et al., "Rapamycin, but Not Resveratrol or Simvastatin, Extends Life Span of Genetically Heterogeneous Mice," *Journals of Gerontology Series A: Biological Sciences and Medical Sciences* 66, no. 2 (February 2011): 191–201.

274 **IN 2014 RAPAMYCIN WAS TESTED AS A LONGEVITY DRUG IN LAB MICE:** Richard A. Miller et al., "Rapamycin-Mediated Lifespan Increase in Mice Is Dose and Sex Dependent and Metabolically Distinct from Dietary Restriction," *Aging Cell* 13, no. 3 (June 2014): 468–77.

274 **TAURINE, AN AMINO ACID FOUND IN MANY FOODS, EXTENDS LIFESPAN:** Parminder Singh et al., "Taurine Deficiency as a Driver of Aging," *Science* 380, no. 6649 (June 9, 2023): eabn9257.

274 **YOU CAN EVEN ARRANGE TO BE TRANSFUSED WITH A YOUNGER PERSON'S BLOOD:** Erin Prater, "Tech CEO Defends Using His 17-Year-Old Son's Blood Plasma in Pursuit of Youth, Despite It Not Working," *Fortune*, July 13, 2023, https://fortune.com/well/2023/07/13/blueprint-ceo-bryan-johnson-defends-plasma-donation-son-youth-aging-longevity-brainstorm-tech-fortune-utah/.

275 **AUSTAD HAS PUBLICLY BET $1 BILLION:** Nic Fleming, "Scientists Up Stakes in Bet on Whether Humans Will Live to 150," *Nature*, October 18, 2016, https://www.nature.com/articles/nature.2016.20818.

276 **FEELING LONELY ACTIVATES YOUR BODY'S FIGHT-OR-FLIGHT STRESS RESPONSE:** John T. Cacioppo et al., "The Neuroendocrinology of Social Isolation," *Annual Review of Psychology* 66 (January 3, 2015): 733–67.

276 **SOCIAL CONNECTEDNESS ADDS YEARS TO LIFE EXPECTANCY:** Fan Wang et al., "A Systematic Review and Meta-Analysis of 90 Cohort Studies of Social Isolation, Loneliness and Mortality," *Nature Human Behaviour* 7, no. 8 (August 2023): 1307–19.

277 **FIGURE 9.1:** O. T. Wolf et al., "The Relationship between Stress Induced Cortisol Levels and Memory Differs between Men and Women," *Psychoneuroendocrinology* 26, no. 7 (October 2001): 711–20.

277 **RACISM ACTIVATES OUR CORTISOL RESPONSE:** Soohyun Nam et al., "Real-Time Racial Discrimination, Affective States, Salivary Cortisol and Alpha-Amylase in Black Adults," *PLoS One* 17, no. 9 (September 14, 2022): e0273081.

277 **IMMIGRANTS CAN SUFFER THE SAME EFFECTS:** Ilona S. Yim et al., "Perceived Stress and Cortisol Reactivity among Immigrants to the United States: The Importance of Bicultural Identity Integration," *Psychoneuroendocrinology* 107 (September 2019): 201–7.

277 **BLUE-COLLAR AND RURAL WHITE COMMUNITIES:** Anne Case and Angus Deaton, "Rising Morbidity and Mortality in Midlife among White Non-Hispanic Americans in the 21st Century," *Proceedings of the National Academy of Science* 112, no. 49 (December 8, 2015): 15078–83.

278 **EASTERN EUROPE WITNESSED A SIMILAR EPIDEMIC IN THE 1990s:** Lawrence King, Gábor Scheiring, and Elias Nosrati, "Deaths of Despair in Comparative Perspective," *Annual Review of Sociology* 48, no. 1 (July 2022): 299–317.

278 **THREE CONSECUTIVE YEARS OF DECREASING IN LIFE EXPECTANCY:** Steven H.

Woolf and Heidi Schoomaker, "Life Expectancy and Mortality Rates in the United States, 1959–2017," *JAMA* 322, no. 20 (November 26, 2019): 1996–2016.

278 **COVID PANDEMIC PUSHED LIFE EXPECTANCY EVEN LOWER IN 2020 AND 2021:** "Life Expectancy in the U.S. Dropped for the Second Year in a Row in 2021," Centers for Disease Control and Prevention, August 31, 2022, https://www.cdc.gov/nchs/pressroom/nchs_press_releases/2022/20220831.htm.

278 **FIGURE 9.2:** Saloni Dattani et al., "Life Expectancy," Our World in Data, accessed February 15, 2024, https://ourworldindata.org/life-expectancy.

INDEX